Multimedia Data Mining

A Systematic Introduction
to Concepts and Theory

Chapman & Hall/CRC
Data Mining and Knowledge Discovery Series

SERIES EDITOR

Vipin Kumar

University of Minnesota
Department of Computer Science and Engineering
Minneapolis, Minnesota, U.S.A

AIMS AND SCOPE

This series aims to capture new developments and applications in data mining and knowledge discovery, while summarizing the computational tools and techniques useful in data analysis. This series encourages the integration of mathematical, statistical, and computational methods and techniques through the publication of a broad range of textbooks, reference works, and handbooks. The inclusion of concrete examples and applications is highly encouraged. The scope of the series includes, but is not limited to, titles in the areas of data mining and knowledge discovery methods and applications, modeling, algorithms, theory and foundations, data and knowledge visualization, data mining systems and tools, and privacy and security issues.

PUBLISHED TITLES

UNDERSTANDING COMPLEX DATASETS: Data Mining with Matrix
Decompositions
David Skillicorn

COMPUTATIONAL METHODS OF FEATURE SELECTION
Huan Liu and Hiroshi Motoda

CONSTRAINED CLUSTERING: Advances in Algorithms, Theory,
and Applications
Sugato Basu, Ian Davidson, and Kiri L. Wagstaff

KNOWLEDGE DISCOVERY FOR COUNTERTERRORISM AND
LAW ENFORCEMENT
David Skillicorn

MULTIMEDIA DATA MINING: A Systematic Introduction to Concepts and Theory
Zhongfei Zhang and Ruofei Zhang

Chapman & Hall/CRC
Data Mining and Knowledge Discovery Series

Multimedia Data Mining

A Systematic Introduction
to Concepts and Theory

Zhongfei Zhang
Ruofei Zhang

CRC Press
Taylor & Francis Group
Boca Raton London New York

CRC Press is an imprint of the
Taylor & Francis Group, an **informa** business
A CHAPMAN & HALL BOOK

The cover images were provided by Yu He, who also participated in the design of the cover page.

CRC Press
Taylor & Francis Group
6000 Broken Sound Parkway NW, Suite 300
Boca Raton, FL 33487-2742

First issued in paperback 2019

© 2009 by Taylor & Francis Group, LLC
CRC Press is an imprint of Taylor & Francis Group, an Informa business

No claim to original U.S. Government works

ISBN-13: 978-1-58488-966-3 (hbk)
ISBN-13: 978-0-367-38627-6 (pbk)

Library of Congress Cataloging-in-Publication Data

Zhang, Zhongfei.
    Multimedia data mining : a systematic introduction to concepts and theory / Zhongfei Zhang, Ruofei Zhang.
        p. cm. -- (Chapman & Hall/CRC data mining and knowledge discovery series)
    Includes bibliographical references and index.
    ISBN 978-1-58488-966-3 (hardcover : alk. paper)
    1. Multimedia systems. 2. Data mining. I. Zhang, Ruofei. II. Title. III. Series.

QA76.575.Z53 2008
006.7--dc22   2008039398

Visit the Taylor & Francis Web site at
http://www.taylorandfrancis.com

and the CRC Press Web site at
http://www.crcpress.com

To my parents, Yukun Zhang and Ming Song; my sister, Xuefei; and my
sons, Henry and Andrew
Zhongfei (Mark) Zhang

To my parents, sister, and wife for their support and tolerance
Ruofei Zhang

Foreword

I am delighted to introduce the first book on multimedia data mining. When I came to know about this book project undertaken by two of the most active young researchers in the field, I was pleased that this book is coming in an early stage of a field that will need it more than most fields do. In most emerging research fields, a book can play a significant role in bringing some maturity to the field. Research fields advance through research papers. In research papers, however, only a limited perspective can be provided about the field, its application potential, and the techniques required and already developed in the field. A book gives such a chance. I liked the idea that there will be a book that will try to unify the field by bringing in disparate topics already available in several papers that are not easy to find and understand. I was supportive of this book project even before I had seen any material on it. The project was a brilliant and a bold idea by two active researchers. Now that I have it on my screen, it appears to be even a better idea.

Multimedia started gaining recognition in the 1990s as a field. Processing, storage, communication, and capture and display technologies had advanced enough that researchers and technologists started building approaches to combine information in multiple types of signals such as audio, images, video, and text. Multimedia computing and communication techniques recognize correlated information in multiple sources as well as insufficiency of information in any individual source. By properly selecting sources to provide complementary information, such systems aspire, much like the human perception system, to create a holistic picture of a situation using only partial information from separate sources.

Data mining is a direct outgrowth of progress in data storage and processing speeds. When it became possible to store large volumes of data and run different statistical computations to explore all possible and even unlikely correlations among data, the field of data mining was born. Data mining allowed people to hypothesize relationships among data entities and explore support for those. This field has been applied to applications in many diverse domains and keeps getting more applications. In fact, many new fields are a direct outgrowth of data mining, and it is likely to become a powerful computational tool behind many emerging natural and social sciences.

Considering the volume of multimedia data and difficulty in developing machine perception systems to bridge the semantic gap, it is natural that multimedia and data mining will come closer and be applied to some of the most challenging problems. And that has started to happen. Some of the toughest challenges for data mining are posed by multimedia systems. Similarly, the potentially most rewarding applications of data mining may come from multimedia data.

As is natural and common, in the early stages of a field people explore only incremental modifications to existing approaches. And multimedia data mining is no exception. Most early tools deal with data in a single medium such as images. This is a good start, but the real challenges are in dealing with multimedia data to address problems that cannot be solved using a single medium. A major limitation of machine perception approaches, so obvious in computer vision but equally common in all other signal based systems, is their over reliance on a single medium. By using multimedia data, one can use an analysis context that is created by a data set of a medium to solve complex problems using data from other media. In a way, multimedia data mining could become a field where analysis will proceed through mutual context propagation approaches. I do hope that some young researchers will be motivated to address these rewarding areas.

This book is the very first monograph on multimedia data mining. The book presents the state-of-the-art materials in the area of multimedia data mining with three distinguishing features. First, this book brings together the literature of multimedia data mining and defines what this area is about, and puts multimedia data mining in perspective compared to other, more well-established research areas.

Second, the book includes an extensive coverage of the foundational theory of multimedia data mining with state-of-the-art materials, ranging from feature extraction and representations, to knowledge representations, to statistical learning theory and soft computing theory. Substantial effort is spent to ensure that the theory and techniques included in the book represent the state-of-the-art research in this area. Though not exhaustive, this book has a comprehensive systematic introduction to the theoretical foundations of multimedia data mining.

Third, in order to showcase to readers the potential and practical applications of the research in multimedia data mining, the book gives specific applications of multimedia data mining theory in order to solve real-world multimedia data mining problems, ranging from image search and mining, to image annotation, to video search and mining, and to audio classification.

While still in its infant stage, multimedia data mining has great momentum to further develop rapidly. It is hoped that the publication of this book shall lead and promote the further development of multimedia data mining research in academia, government, and industries, and its applications in all the sectors of our society.

Ramesh Jain
University of California at Irvine

About the Authors

Zhongfei (Mark) Zhang is an associate professor in the Computer Science Department at the State University of New York (SUNY) at Binghamton, and the director of the Multimedia Research Laboratory in the Department. He received a BS in Electronics Engineering (with Honors), an MS in Information Sciences, both from Zhejiang University, China, and a PhD in Computer Science from the University of Massachusetts at Amherst. He was on the faculty of the Computer Science and Engineering Department, and a research scientist at the Center of Excellence for Document Analysis and Recognition, both at SUNY Buffalo. His research interests include multimedia information indexing and retrieval, data mining and knowledge discovery, computer vision and image understanding, pattern recognition, and bioinformatics. He has been a principal investigator or co-principal investigator for many projects in these areas supported by the US federal government, the New York State government, as well as private industries. He holds many inventions, has served as a reviewer or a program committee member for many conferences and journals, has been a grant review panelist every year since 2000 for the federal government funding agencies (mainly NSF and NASA), New York State government funding agencies, and private funding agencies, and has served on the editorial board for several journals. He has also served as a technical consultant for a number of industrial and governmental organizations and is a recipient of several prestigious awards.

Ruofei Zhang is a computer scientist and technical manager at Yahoo! Inc. He has led the relevance R&D in Yahoo! Video Search and the contextual advertising relevance modeling and optimization group in Search & Advertising Science at Yahoo!. When he was in graduate school, he worked as a research intern at Microsoft Research Asia. His research fields are in machine learning, large scale data analysis and mining, optimization, and multimedia information retrieval. He has published over two dozen peer-reviewed academic papers in leading international journals and conferences, has written several invited papers and book chapters, has filed 10 patents on search relevance, ranking function learning, multimedia content analysis, and has served as a reviewer or a program committee member for many prestigious international journals and conferences. He is a Member of IEEE, a member of the IEEE Computer Society, and a member of ACM. He received a PhD in Computer Science with a Distinguished Dissertation Award from the State University of New York at Binghamton.

Contents

List of Tables

List of Figures

Preface

Multimedia data mining is a very interdisciplinary and multidisciplinary area. This area was developed under the two parent areas — *multimedia* and *data mining*. Since both parent areas are considered young areas with the history of around the last ten years or so, the formal development of multimedia data mining was not even established until very recently. This book is the very first monograph in the general area of multimedia data mining written in a self-contained format. This book addresses both the fundamentals and the applications of multimedia data mining. It gives a systematic introduction to the fundamental theory and concepts in this area, and at the same time, also presents specific applications that showcase the great potential and impacts for the technologies generated from the research in this area.

The authors of this book have been actively working in this area for years, and this book is the final culmination of their years of long research in this area. This book may be used as a collection of research notes for researchers in this area, a reference book for practitioners or engineers, as well as a textbook for a graduate advanced seminar in this area or any related areas. This book may also be used for an introductory course for graduate students or advanced undergraduate seniors. The references collected in this book may be used as further reading lists or references for the readers.

Due to the very interdisciplinary and multidisciplinary nature of the area of multimedia data mining, and also due to the rapid development in this area in the recent years, it is by no means meant to be exhaustive to collect complete information in this area. We have tried our best to collect the most recent developments related to the specific topics addressed in this book in the general area of multimedia data mining. For those who have already been in the area of multimedia data mining or who already know what this area is about, this book serves the purpose of a formal and systematic collection to connect all of the dots together. For those who are beginners to the area of multimedia data mining, this book serves the purpose of a formal and systematic introduction to this area.

It is not possible for us to accomplish this book without the great support from a large group of people and organizations. In particular, we would like to thank the publisher — Taylor & Francis/CRC Press for giving us the opportunity to complete this book for the readers as one of the books in the Chapman & Hall/CRC *Data Mining and Knowledge Discovery* series, with Prof. Vipin Kumar at the University of Minnesota serving as the series editor. We would like to thank this book's editor of Taylor & Francis Group, Randi Cohen, for her enthusiastic and patient support, effort, and advice; the project editor of Taylor & Francis Group, Judith M. Simon, and the anonymous proof-reader for their meticulous effort in correcting typos and other errors of the draft of the book; and Shashi Kumar of International Typesetting and Composition for his prompt technical support in formatting the book. We would like to thank Prof. Ramesh Jain at the University of California at Irvine for the strong support to this book and kindly offering to write a foreword to this book. We would like to thank Prof. Ying Wu at Northwestern University and Prof. Chabane Djeraba at the University of Science and Technology of Lille, France, as well as another anonymous reviewer, for their painstaking effort to review the book and their valuable comments to substantially improve the quality of this book. Part of the book is derived from the original contributions made by the authors of the book as well as a group of their colleagues. We would like to specifically thank the following colleagues for their contributions: Jyh-Herng Chow, Wei Dai, Alberto del Bimbo, Christos Faloutsos, Zhen Guo, Ramesh Jain, Mingjing Li, Wei-Ying Ma, Florent Masseglia, Jia-Yu (Tim) Pan, Ramesh Sarukkai, Eric P. Xing, and HongJiang Zhang. This book project is supported in part by the National Science Foundation under grant IIS-0535162, managed by the program manager, Dr. Maria Zemankova. Any opinions, findings, and conclusions or recommendations expressed in this material are those of the authors and do not necessarily reflect the views of the National Science Foundation.

Finally, we would like to thank our families for the love and support that are essential for us to complete this book.

Part I

Introduction

Chapter 1

Introduction

1.1 Defining the Area

Multimedia data mining, as the name suggests, presumably is a combination of the two emerging areas: *multimedia* and *data mining*. However, multimedia data mining is *not* a research area that just simply combines the research of multimedia and data mining together. Instead, the multimedia data mining research focuses on the theme of merging multimedia and data mining research together to exploit the synergy between the two areas to promote the understanding and to advance the development of the knowledge discovery in multimedia data. Consequently, multimedia data mining exhibits itself as a unique and distinct research area that synergistically relies on the state-of-the-art research in multimedia and data mining but at the same time fundamentally differs from either multimedia or data mining or a simple combination of the two areas.

Multimedia and data mining are two very interdisciplinary and multidisciplinary areas. Both areas started in early 1990s with only a very short history. Therefore, both areas are relatively young areas (in comparison, for example, with many well established areas in computer science such as operating systems, programming languages, and artificial intelligence). On the other hand, with substantial application demands, both areas have undergone independently and simultaneously rapid developments in recent years.

Multimedia is a very diverse, interdisciplinary, and multidisciplinary research area[1]. The word *multimedia* refers to a combination of multiple media types together. Due to the advanced development of the computer and digital technologies in early 1990s, multimedia began to emerge as a research area [87, 197]. As a research area, multimedia refers to the study and development of an effective and efficient multimedia system targeting a specific application. In this regard, the research in multimedia covers a very wide spectrum of subjects, ranging from multimedia indexing and retrieval, multimedia databases, multimedia networks, multimedia presentation, multimedia

---

[1]Here we are only concerned with a research area; multimedia may also be referred to industries and even social or societal activities.

quality of services, multimedia usage and user study, to multimedia standards, just to name a few.

While the area of multimedia is so diverse with many different subjects, those that are related to multimedia data mining mainly include multimedia indexing and retrieval, multimedia databases, and multimedia presentation [72, 113, 198]. Today, it is well known that multimedia information is ubiquitous and is often required, if not necessarily essential, in many applications. This phenomenon has made multimedia repositories widespread and extremely large. There are tools for managing and searching within these collections, but the need for tools to extract hidden useful knowledge embedded within multimedia collections is becoming pressing and central for many decision-making applications. For example, it is highly desirable for developing the tools needed today for discovering relationships between objects or segments within images, classifying images based on their content, extracting patterns in sound, categorizing speech and music, and recognizing and tracking objects in video streams.

At the same time, researchers in multimedia information systems, in the search for techniques for improving the indexing and retrieval of multimedia information, are looking for new methods for discovering indexing information. A variety of techniques, from machine learning, statistics, databases, knowledge acquisition, data visualization, image analysis, high performance computing, and knowledge-based systems, have been used mainly as research handcraft activities. The development of multimedia databases and their query interfaces recalls again the idea of incorporating multimedia data mining methods for dynamic indexing.

On the other hand, data mining is also a very diverse, interdisciplinary, and multidisciplinary research area. The terminology *data mining* refers to knowledge discovery. Originally, this area began with knowledge discovery in databases. However, data mining research today has been advanced far beyond the area of databases [71, 97]. This is due to the following two reasons. First, today's knowledge discovery research requires more than ever the advanced tools and theory beyond the traditional database area, noticeably mathematics, statistics, machine learning, and pattern recognition. Second, with the fast explosion of the data storage scale and the presence of multimedia data almost everywhere, it is not enough for today's knowledge discovery research to just focus on the structured data in the traditional databases; instead, it is common to see that the traditional databases have evolved into data warehouses, and the traditional structured data have evolved into more non-structured data such as imagery data, time-series data, spatial data, video data, audio data, and more general multimedia data. Adding into this complexity is the fact that in many applications these non-structured data do not even exist in a more traditional "database" anymore; they are just simply a collection of the data, even though many times people still call them databases (e.g., image database, video database).

Examples are the data collected in fields such as art, design, hypermedia and digital media production, case-based reasoning and computational modeling of creativity, including evolutionary computation, and medical multimedia data. These exotic fields use a variety of data sources and structures, interrelated by the nature of the phenomenon that these structures describe. As a result there is an increasing interest in new techniques and tools that can detect and discover patterns that lead to new knowledge in the problem domain where the data have been collected. There is also an increasing interest in the analysis of multimedia data generated by different distributed applications, such as collaborative virtual environments, virtual communities, and multi-agent systems. The data collected from such environments include a record of the actions in them, a variety of documents that are part of the business process, asynchronous threaded discussions, transcripts from synchronous communications, and other data records. These heterogeneous multimedia data records require sophisticated preprocessing, synchronization, and other transformation procedures before even moving to the analysis stage.

Consequently, with the independent and advanced developments of the two areas of multimedia and data mining, with today's explosion of the data scale and the existence of the pluralism of the data media types, it is natural to evolve into this new area called *multimedia data mining*. While it is presumably true that multimedia data mining is a combination of the research between multimedia and data mining, the research in multimedia data mining refers to the synergistic application of knowledge discovery theory and techniques in a multimedia database or collection. As a result, "inherited" from its two parent areas of multimedia and data mining, multimedia data mining by nature is also an interdisciplinary and multidisciplinary area; in addition to the two parent areas, multimedia data mining also relies on the research from many other areas, noticeably from mathematics, statistics, machine learning, computer vision, and pattern recognition. Figure 1.1 illustrates the relationships among these interconnected areas.

While we have clearly given the working definition of multimedia data mining as an emerging, active research area, due to historic reasons, it is helpful to clarify several misconceptions and to point out several pitfalls at the beginning.

- *Multimedia Indexing and Retrieval* vs. *Multimedia Data Mining*: It is well-known that in the classic data mining research, the pure text retrieval or the classic information retrieval is *not* considered as part of data mining, as there is no knowledge discovery involved. However, in multimedia data mining, when it comes to the scenarios of multimedia indexing and retrieval, this boundary becomes vague. The reason is that a typical multimedia indexing and/or retrieval system reported in the recent literature often contains a certain level of knowledge discovery such as feature selection, dimensionality reduction, concept discovery, as well as mapping discovery between different modalities (e.g., imagery annotation where a mapping from an image to textual words is discov-

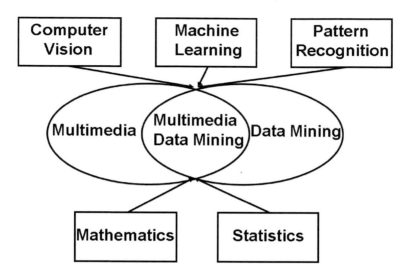

FIGURE 1.1: Relationships among the interconnected areas to multimedia data mining.

ered and word-to-image retrieval where a mapping from a textual word to images is discovered). In this case, multimedia information indexing and/or retrieval is considered as part of multimedia data mining. On the other hand, if a multimedia indexing or retrieval system uses a "pure" indexing system such as the text-based indexing technology employed in many commercial imagery/video/audio retrieval systems on the Web, this system is not considered as a multimedia data mining system.

- *Database* vs. *Data Collection*: In a classic database system, there is always a database management system to govern all the data in the database. This is true for the classic, structured data in the traditional databases. However, when the data become non-structured data, in particular, multimedia data, often we do not have such a management system to "govern" all the data in the collection. Typically, we simply just have a whole collection of multimedia data, and we expect to develop an indexing/retrieval system or other data mining system on top of this data collection. For historic reasons, in many literature references, we still use the terminology of "database" to refer to such a multimedia data collection, even though this is different from the traditional, structured database in concept.

- *Multimedia Data* vs. *Single Modality Data*: Although "multimedia" refers to the multiple modalities and/or multiple media types of data, conventionally in the area of multimedia, multimedia indexing and retrieval also includes the indexing and retrieval of a single, non-text modality of data, such as image indexing and retrieval, video indexing and retrieval, and audio indexing and retrieval. Consequently, in multimedia data mining, we follow this convention to include the study of any knowledge discovery dedicated to any single modality of data as part of the multimedia data mining research. Therefore, studies in image data mining, video data mining, and audio data mining alone are considered as part of the multimedia data mining area.

Multimedia data mining, although still in its early booming stage as an area that is expected to have further development, has already found enormous application potential in a wide spectrum covering almost all the sectors of society, ranging from people's daily lives to economic development to government services. This is due to the fact that in today's society almost all the real-world applications often have data with multiple modalities, from multiple sources, and in multiple formats. For example, in homeland security applications, we may need to mine data from an air traveler's credit history, traveling patterns, photo pictures, and video data from surveillance cameras in the airport. In the manufacturing domains, business processes can be improved if, for example, part drawings, part descriptions, and part flow can be mined in an integrated way instead of separately. In medicine, a disease might be predicted more accurately if the MRI (magnetic resonance imaging) imagery is mined together with other information about the patient's condition. Similarly, in bioinformatics, data are available in multiple formats.

1.2 A Typical Architecture of a Multimedia Data Mining System

A typical multimedia data mining system, or framework, or method always consists of the following three key components. Given the raw multimedia data, the very first step for mining the multimedia data is to convert a specific raw data collection (or a database) into a representation in an abstract space which is called the feature space. This process is called feature extraction. Consequently, we need a feature representation method to convert the raw multimedia data to the features in the feature space, before any mining activities are able to be conducted. This component is very important as the success of a multimedia data mining system to a large degree depends upon how good the feature representation method is. The typical feature representation methods or techniques are taken from the classic computer vision research, pattern recognition research, as well as multimedia information indexing and retrieval research in multimedia area.

Since knowledge discovery is an intelligent activity, like other types of intelligent activities, multimedia data mining requires the support of a certain level of knowledge. Therefore, the second key component is the knowledge representation, i.e., how to effectively represent the required knowledge to support the expected knowledge discovery activities in a multimedia database. The typical knowledge representation methods used in the multimedia data mining literature are directly taken from the general knowledge representation research in artificial intelligence area with the possible special consideration in the multimedia data mining problems such as spatial constraints based reasoning.

Finally, we come to the last key component — the actual mining or learning theory and/or technique to be used for the knowledge discovery in a multimedia database. In the current literature of multimedia data mining, there are mainly two paradigms of the learning or mining theory/techniques that can be used separately or jointly in a specific multimedia data mining application. They are *statistical learning theory* and *soft computing theory*, respectively. The former is based on the recent literature on machine learning and in particular statistical machine learning, whereas the latter is based on the recent literature on soft computing such as fuzzy logic theory. This component typically is the core of the multimedia data mining system.

In addition to the three key components, in many multimedia data mining systems, there are user interfaces to facilitate the communications between the users and the mining systems. Like the general data mining systems, for a typical multimedia data mining system, the quality of the final mining results can only be judged by the users. Hence, it is necessary in many cases to have a user interface to allow the communications between the users and the mining systems and the evaluations of the final mining quality; if the quality is not acceptable, the users may need to use the interface to tune different parameter values of a specific component used in the system, or even to change different components, in order to achieve better mining results, which may go into an iterative process until the users are happy with the mining results.

Figure 1.2 illustrates this typical architecture of a multimedia data mining system.

1.3   The Content and the Organization of This Book

This book aims at defining the area of multimedia data mining. We give a systematic introduction to this area by outlining what this area is about, what is considered as the theory of this area, and what are the examples of the applications of multimedia data mining. Since this area is so diverse, interdisciplinary, and multidisciplinary, this introduction as well as the materials covered in this book can by no means be exhaustive and complete. We have tried our best to select materials included in this book that are representative

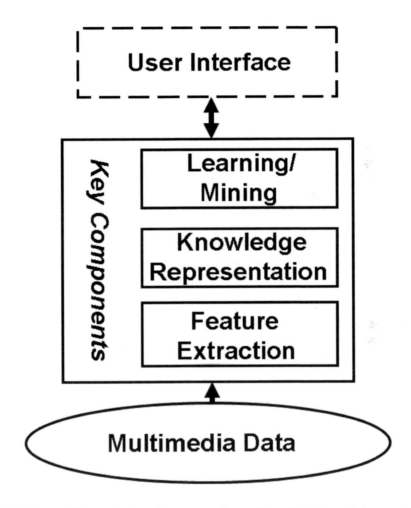

FIGURE 1.2: The typical architecture of a multimedia data mining system.

enough to expose the readers to the whole area of multimedia data mining as much as possible under the limited time constraint to publish this book. On the other hand, due to the rapid development in the literature of this area, we have also tried our best to select the materials that represent the most recent advances and status quo of the development of multimedia data mining.

The organization of this book is as follows. The whole book contains three parts. Part I is this Introduction chapter to define the area of multimedia data mining and to outline what this book is about. Part II is dedicated to the theoretical foundation of the area of multimedia data mining. Specifically, there are three chapters in this Part. Chapter 2 introduces the commonly used feature representation techniques and the knowledge representation techniques in multimedia data mining research. Chapter 3 introduces the commonly used statistical theory and techniques for multimedia data mining. Chapter 4 introduces the commonly used soft computing theory and techniques for multimedia data mining. Finally, Part III showcases application examples in multimedia data mining research. Specifically, there are five chapters in this Part. Chapter 5 presents an image database modeling approach to multimedia data mining; the focus is to develop a semantic repository training method. Chapter 6 presents another image database modeling approach to multimedia data mining where the focus is on developing a concept discovery method in an imagery database. Chapter 7 presents yet another example in imagery data mining where we address a specific image mining problem — imagery annotation, in which we demonstrate how knowledge discovery helps achieve the goal of imagery annotation. Chapter 8 demonstrates the application of video data mining to developing an effective solution to large-scale video search on the Web. Chapter 9 describes an application of audio data classification and categorization.

1.4   The Audience of This Book

This book is a monograph on the authors' recent research in the emerging area of multimedia data mining. Therefore, the expected readership of this book is all the researchers and system developing engineers working in the area of multimedia data mining as well as all the related areas, including but not limited to, multimedia, data mining, machine learning, computer vision, pattern recognition, statistics, as well as other application areas that use multimedia data mining techniques such as bioinformatics and marketing. Since this book is self-contained in the presentations of the materials, this book also serves as an ideal reference book for people who are interested in the new area of multimedia data mining. Consequently, in addition, the readership also includes any of those who have this interest or work in a field which needs this reference book. Finally, this book can be used as a textbook for a graduate course or even undergraduate senior elective course on the topic of multimedia data mining, as it provides a systematic introduction to this area.

1.5 Further Readings

As is defined in Section 1.1, the area of multimedia data mining emerges from the two independent areas of multimedia and data mining. Therefore, the history of multimedia data mining may trace back to the histories of the two parent areas. Since multimedia data mining is just in its infant stage, currently there is no dedicated, premier venue for the publications of the research in this area. Consequently, the related work in this area, as the supplementary information to this book for further readings, may be found in the literature of the two parent areas. Specifically, in the multimedia area, related work may be found in the premier conferences of ACM Multimedia (ACM MM) and IEEE International Conference on Multimedia and Expo (IEEE ICME). In particular, the most relevant venue is the annual ACM International Conference on Multimedia Information Retrieval (ACM MIR), which used to be an annual workshop in conjunction with ACM MM. Also recently, there has been a new premier conference that is dedicated to image and video retrieval, ACM International Conference on Image and Video Retrieval (ACM CIVR). In addition, much of the related work may be found in the computer vision premier conferences, noticeably, IEEE International Conference on Computer Vision (IEEE ICCV), IEEE International Conference on Computer Vision and Pattern Recognition (IEEE CVPR), and European Conference on Computer Vision (ECCV). Some of the related work may also be found in the pattern recognition premier conference, International Conference on Pattern Recognition (ICPR), as well as the audio and speech signal processing premier conference, International Conference on Audio and Speech Signal Processing (ICASSP). For journals, the related work may be found in the premier journals in the multimedia area as well as the related areas of computer vision and pattern recognition, including *IEEE Transactions on Multimedia* (IEEE T-MM), *IEEE Transactions on Pattern Analysis and Machine Intelligence* (IEEE T-PAMI), *IEEE Transactions on Image Processing* (IEEE TIP), *IEEE Transactions on Speech and Audio Processing* (IEEE T-SAP), and *Pattern Recognition* (PR), as well as the recently inaugurated journal, *ACM Transactions on Multimedia Computing, Communications and Applications* (ACM TOMCCAP).

In the data mining area, related work may be found in the premier conferences such as ACM International Conference on Knowledge Discovery and Data Mining (ACM KDD), IEEE International Conference on Data Mining

(IEEE ICDM), and SIAM International Conference on Data Mining (SDM). In particular, related work may be found in the annual workshop dedicated to the area of multimedia data mining in conjunction with the annual ACM KDD conference, the International Workshop on Multimedia Data Mining (ACM MDM). For journals, the premier journals in the data mining area may contain related work in multimedia data mining, including *IEEE Transactions on Knowledge and Data Engineering* (IEEE T-KDE) and *ACM Transactions on Data Mining* (ACM TDM).

In addition, there is recently published literature that is related to multimedia data mining. Gong and Xu [90] introduce the machine learning techniques commonly used in multimedia research. Petrushin and Khan [167]; Zaiane, Smirof, and Djeraba [233]; and Djeraba [62] edited collections on the recent research in the area of multimedia data mining. Zhang et al [250] edited a special issue in *IEEE Transactions on Multimedia* dedicated to the area of multimedia data mining.

Part II

Theory and Techniques

Chapter 2

Feature and Knowledge Representation for Multimedia Data

2.1 Introduction

Before we study multimedia data mining, the very first issue we must resolve is how to represent multimedia data. While we can always represent the multimedia data in their original, raw formats (e.g., imagery data in their original formats such as JPEG, TIFF, or even the raw matrix representation), due to the following two reasons, these original formats are considered as awkward representations in a multimedia data mining system, and thus are rarely used directly in any multimedia data mining applications.

First, these original formats typically take much more space than necessary. This immediately poses two problems – more processing time and more storage space. Second and more importantly, these original formats are designed for best archiving the data (e.g., for minimally losing the integrity of the data while at the same time for best saving the storage space), but not for best fulfilling the multimedia data mining purpose. Consequently, what these original formats have represented are just the *data*. On the other hand, for the multimedia data mining purpose, we intend to represent the multimedia data as useful *information* that would facilitate different processing and mining operations. For example, Figure 2.1(a) shows an image of a horse. For such an image, the original format is in JPEG and the actual "content" of this image is the binary numbers for each byte in the original representation which does not tell anything about what this image is. Ideally, we would expect the representation of this image as the useful information such as the way represented in Figure 2.1(b). This representation would make the multimedia data mining extremely easy and straightforward.

However, this immediately poses a chicken-and-egg problem – the goal of the multimedia data mining is to discover the knowledge represented in an appropriate way, whereas if we were able to represent the multimedia data in such a concise and semantic way as shown in the example in Figure 2.1(b), the problem of multimedia data mining would already have been solved. Consequently, as a "compromise", instead of directly representing the multimedia data in a semantic knowledge representation such as that in Figure 2.1(b), we

 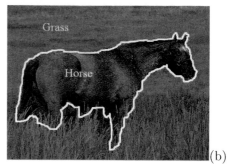

(a)   (b)

FIGURE 2.1: (a) An original image; (b) An ideal representation of the image in terms of the semantic content.

first represent the multimedia data as *features*. In addition, in order to effectively mine the multimedia data, in many multimedia data mining systems, additional knowledge representation is also used to appropriately represent different types of knowledge associated with the multimedia data for the mining purpose, such as domain knowledge, background knowledge, and common sense knowledge.

The rest of this chapter is organized as follows. While the feature and knowledge representation techniques introduced in this chapter are applicable to all the different media types and/or modalities, we first introduce several commonly used concepts in multimedia data mining, and some of them are media-specific concepts, at the very beginning of this chapter, in Section 2.2. Section 2.3 then introduces the commonly used features for multimedia data, including statistical features, geometric features, and meta features. Section 2.4 introduces the commonly used knowledge representation methods in the multimedia data mining applications, including logic based representation, semantic networks based representation, frame based representation, as well as constraint based representation; we also introduce the representation methods on uncertainty. Finally, this chapter is concluded in Section 2.5.

2.2   Basic Concepts

Before we introduce the commonly used feature and knowledge representation techniques that are typically applicable to all the media types and/or modalities of data, we begin with introducing several important and commonly used concepts related to multimedia data mining. Some of these concepts are applicable to all the media types, while others are media-specific.

2.2.1 Digital Sampling

While multimedia data mining, like its parent areas of data mining and multimedia, essentially deals with digital representations of the information through computers, the world we live with is actually in a continuous space. Most of the time, what we see is a continuous scene; what we hear is continuous sound (music, human talking, many of the environmental sounds, or even many of the noises such as a vehicle horn beep). The only exception is probably what we read, which are the words that consist of characters or letters that are sort of digital representations. In order to transform the continuous world into a digital representation that a computer can handle, we need to *digitize* or *discretize* the original continuous information to the digital representations known to a computer as data. This digitization or discretization process is performed through *sampling*.

There are three types of sampling that are needed to transform the continuous information to the digital data representations. The first type of sampling is called *spatial sampling*, which is for the spatial signals such as imagery. Figure 2.2(a) shows the spatial sampling concept. For imagery data, each sample obtained after the spatial sampling is called a *pixel*, which stands for a picture element. The second type of sampling is called *temporal sampling*, which is for the temporal signals such as audio sounds. Figure 2.2(b) shows the temporal sampling concept. For audio data, after the temporal sampling, a fixed number of neighboring samples along the temporal domain is called a *frame*. Typically, in order to exploit the temporal redundancy for certain applications such as compression, it is intentionally left as an overlap between two neighboring frames for at least one third of a frame-size.

For certain continuous information such as video signals, both spatial and temporal samplings are required. For the video signals, after the temporal sampling, a continuous video becomes a sequence of temporal samples, and now each such temporal sample becomes an image, which is called a *frame*. Each frame, since it is actually an image, can be further spatially sampled to have a collection of pixels. For video data, in each frame, it is common to define a fixed number of spatially contiguous pixels as a *block*. For example, in the MPEG format [4], a block is defined as a region of 8×8 pixels.

Temporal data such as audio or video are often called *stream* data. Stream data can be cut into exclusive segments along the temporal axis. These segments are called *clips*. Thus, we have video clip files or audio clip files.

Both the spatial sampling and the temporal sampling must follow a certain rule in order to ensure that the sampled data reflect the original continuous information without losing anything. Clearly, this is important as under-sampling shall lose essential information and over-sampling shall generate more data than necessarily required. The optimal sampling frequency is shown to be the twice the highest structural change frequency (for spatial sampling) or twice the highest temporal change frequency (for temporal sampling). This rule is called the *Nyquist Sampling Theorem* [160], and this

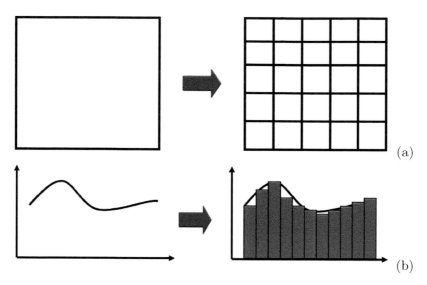

FIGURE 2.2: (a) A spatial sampling example. (b) A temporal sampling example.

optimal sampling frequency is called the *Nyquist frequency*.

The third type of sampling is called *signal sampling*. After the spatial or temporal sampling, we have a collection of samples. The actual measuring space of these samples is still continuous. For example, after a continuous image is spatially sampled into a collection of samples, these samples represent the brightness values at the different sampling locations of the image, and the brightness is a continuous space. Therefore, we need to apply the third type of sampling, the signal sampling, to the brightness space to represent a continuous range of the original brightness into a finite set of digital signal values. This is what the signal sampling is for. Depending upon different application needs, the signal sampling may follow a linear mathematical model (such as that shown in Figure 2.3(a)) or a non-linear mathematical model (such as that shown in Figure 2.3(b)).

2.2.2   Media Types

From the conventional database terminologies, all the data that can be represented and stored in the conventional database structures, including the commonly used relational database and object-oriented database structures, are called *structured data*. Multimedia data, on the other hand, often refer to the data that cannot be represented or indexed in the conventional database structures and, thus, are often called *non-structured data*. Non-structured data can then be further defined in terms of the specific media types they be-

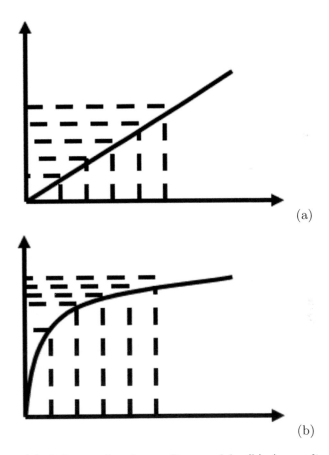

FIGURE 2.3: (a) A linear signal sampling model. (b) A non-linear signal sampling model.

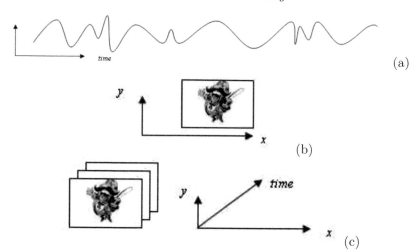

FIGURE 2.4: (a) One-dimensional media type data. (b) Two-dimensional media type data. (c) Three-dimensional media type data.

long to. There are several commonly encountered media types in multimedia data mining. They can be represented in terms of the dimensions of the space the data are in. Specifically, we list those commonly encountered media types as follows.

- *0-dimensional data*: This type of the data is the regular, alphanumeric data. A typical example is the text data.

- *1-dimensional data*: This type of the data has one dimension of a space imposed into them. A typical example of this type of the data is the audio data, as shown in Figure 2.4(a).

- *2-dimensional data*: This type of the data has two dimensions of a space imposed into them. Imagery data and graphics data are the two common examples of this type of data, as shown in Figure 2.4(b).

- *3-dimensional data*: This type of the data has three dimensions of a space imposed into them. Video data and animation data are the two common examples of this type of data, as shown in Figure 2.4(c).

As introduced in Chapter 1, the very first things for multimedia data mining are the feature extraction and knowledge representation. While there are many feature and knowledge representation techniques that are applicable to all different media types, as are introduced in the rest of this chapter, there are several media-specific feature representations that we briefly introduce below.

- *TF-IDF*: The TF-IDF measure is specifically defined as a feature for text data. Given a text database of N documents and a total M word vocabulary, the standard text processing model is based on the *bag-of-words* assumption, which says that for all the documents, we do not consider any linguistic or spatial relationship between the words in a document; instead, we consider each document just as a collection of isolated words, resulting in a bag-of-words representation. Given this assumption, we represent the database as an $N \times M$ matrix which is called the *Term Frequency Matrix*, where each entry $TF(i, j)$ is the occurrence frequency of the word j occurring in the document i. Therefore, the total term frequency for the word j is

$$TF(j) = \sum_{i=1}^{N} TF(i, j) \qquad (2.1)$$

In order to penalize those words that appear too frequently, which does not help in indexing the documents, an *inverse document frequency* (IDF) is defined as

$$IDF(j) = \log \frac{N}{DF(j)} \qquad (2.2)$$

where $DF(j)$ means the number of the documents in which the word j appears, and is called the *document frequency* for the word j. Finally, TF-IDF for a word j is defined as

$$\text{TF-IDF}(j) = TF(j) \times IDF(j) \qquad (2.3)$$

The details of the TF-IDF feature may be found in [184].

- *Cepstrum*: Cepstrum features are often used for one-dimensional media type data such as audio data. Given such a media type data represented as a one-dimensional signal, *cepstrum* is defined as the Fourier transform of the signal's decibel spectrum. Since the decibel spectrum of a signal is obtained by taking the logarithm of the Fourier transform of the original signal, cepstrum is sometimes in the literature also called the *spectrum of a spectrum*. The technical details of the cepstral features may be found in [49].

- *Fundamental Frequency*: This refers to the lowest frequency in a series of harmonics a typical audio sound has. If we represent the audio sound in terms of a series of sinusoidal functions, the fundamental frequency refers to the frequency that the sinusoidal function with the lowest frequency in the spectrum has. Fundamental frequency is often used as a feature for audio data mining.

- *Audio Sound Attributes*: Typical audio sound attributes include pitch, loudness, and timbre. *Pitch* refers to the sensation of the "altitude" or the "height", often related to the frequency of the sounds, in particular, related to the fundamental frequency of the sounds. *Loudness* refers to the sensation of the "strength" or the "intensity" of the sound tone, often related to the sound energy intensity (i.e., the energy flow or the oscillation amplitude of the sound wave reaching the human ear). *Timbre* refers to the sensation of the "quality" of the audio sounds, often related to the spectrum of the audio sounds. The details of these attributes may be found in [197]. These attributes are often used as part of the features for audio data mining.

- *Optical Flow*: Optical flows are the features often used for three-dimensional media type data such as video and animation. Optical flows are defined as the changes of an image's brightness of a specific location of an image over the time in the motion pictures such as video or animation streams. A related but different concept is called *motion field*, which is defined as the motion of a physical object in a three-dimensional space measured at a specific point on the surface of this object mapped to a corresponding point in a two-dimensional image over the time. Motion vectors are useful information in recovering the three-dimensional motion from an image sequence in computer vision research [115]. Since there is no direct way to measure the motion vectors in an image plane, often it is assumed that the motion vectors are the same as the optical flows and thus the optical flows are used as the motion vectors. However, conceptually they are different. For the details of the optical flows as well as their relationship to the motion vectors, see [105].

2.3 Feature Representation

Given a specific modality of the multimedia data (e.g., imagery, audio, and video), feature extraction is typically the very first step for processing and mining. In general, features are the abstraction of the data in a specific modality defined in measurable quantities in a specific Euclidean space [86]. The Euclidean space is thus called *feature space*. Features, also called *attributes*, are an abstract description of the original multimedia data in the feature space. Since typically there are more than one feature used to describe the data, these multiple features form a *feature vector* in the feature space. The process of identifying the feature vector from the original multimedia data is called *feature extraction*. Depending upon different features defined in a multimedia system, different feature extraction methods are used to obtain these features.

Typically, features are defined with respect to a specific modality of the multimedia data. Consequently, given multiple modalities of multimedia data, we may use a feature vector to describe the data in each modality. As a result, we may use a combined feature vector for all the different modalities of the data (e.g., a concatenation of all the feature vectors for different modalities) if the mining is to be performed in the whole data collection aggregatively, or we may leave the individual feature vectors for the individual modalities of the data if the mining is to be performed for different modalities of the data separately.

Essentially, there are three categories of features that are often used in the literature. They are *statistical features*, *geometric features*, and *meta features*. Except for some of the meta features, most of the feature representation methods are applied to a *unit* of multimedia data instead of to the whole multimedia data, or even to a part of a multimedia data unit. A unit of multimedia data is typically defined with respect to a specific modality of the data. For example, for an audio stream, a unit is an audio frame; for an imagery collection, a unit is an image; for a video stream, a unit is a video frame. A part of a multimedia data unit is called an *object*. An object is obtained by a segmentation of the multimedia data unit. In this sense, the feature extraction is a mapping from a multimedia data unit or an object to a feature vector in a feature space. We say that a feature is *unique* if and only if different multimedia data units or different objects map to different values of the feature; in other words, the mapping is one-to-one. However, when this uniqueness definition of features is carried out to the object level instead of the multimedia data unit level, different objects are interpreted in terms of different *semantic objects* as opposed to different variations of the same object. For example, an apple and an orange are two different semantic objects, while different views of the same apple are different variations of the same object but not different semantic objects.

In this section, we review several well-known feature representation methods in each of the categories.

2.3.1 Statistical Features

Statistical features focus on a statistical description of the original multimedia data in terms of a specific aspect such as the frequency counts for each of the values of a specific quantity of the data. Consequently, all the statistical features only give an aggregate, statistical description of the original data in an *aspect*, and therefore, it is in general not possible to expect to recover the original information from this aggregate, statistical description. In other words, statistical features are typically *not* unique; if we conceptualize obtaining the statistical features from the original data as a transformation, this transformation is, in general, lossy. Unlike geometric features, statistical features are typically applied to the whole multimedia data unit without segmentation of the unit into identified parts (such as an object) instead of to the parts. Due to this reason, in general all the variation-invariant properties (e.g., translation-invariant, rotation-invariant, scale-invariant, or the more general affine-invariant) for any segmented part of a multimedia data unit do not hold true for statistical features.

Well-known statistical features include histograms, transformation coefficients, coherent vectors, and correlograms. We give a brief review of each of these statistical features.

2.3.1.1 Histogram

Histograms as a well-known feature date back to the early literature of pattern recognition and image analysis [67]. A histogram is a statistical method to convert an original data representation to the *occurrence frequency* information measured for a specific quantity in the original data; consequently, a histogram is represented as a 1-dimensional vector, where the X-axis is the range of this specific quantity and the Y-axis is the occurrence frequency information measured for each value in the range of the specific quantity. The specific quantity depends on different data modalities and also on different applications, and is often defined in advance by the user. For example, given an image, we may use the image intensity value as the specific quantity; we may also use the image optical flow magnitude as the specific quantity.

Mathematically, given a specific quantity $F(\boldsymbol{p})$ as a function of a sample vector $\boldsymbol{p}$ of the multimedia data (e.g., $\boldsymbol{p}$ may be a spatial point represented as a pair of coordinates $\boldsymbol{p} = (x, y)^T$ for an image), the histogram $H(r)$ defined with respect to this quantity $F(\boldsymbol{p})$ for a value r in the range R of the function $F(\boldsymbol{p})$ is defined as follows:

$$\forall r \in R, H(r) = \sum_{\forall \boldsymbol{p}} \delta(F(\boldsymbol{p}) = r) \qquad (2.4)$$

where δ is the Kronecker Delta function. In the definition Equation 2.4, it is assumed that the whole domain of the variable r is quantized into a parameter of the granularity b. This parameter b is called *bucket size*. With a predetermined bucket size b, a histogram $H(r)$ is a vector, with the dimensionality of the vector depending upon the specific value of b. A larger bucket size results in a "coarser" histogram with a lower dimensionality, while a smaller bucket size results in a "finer" histogram with a higher dimensionality. For example, given an image, assuming that the intensity is quantized to the range $[0, 255]$, if $b = 1$, the histogram has 256 buckets and the dimensionality is 256; if $b = 10$, the histogram has 26 buckets and the dimensionality is 26. Figure 2.5(a) illustrates a small image represented in the original intensity values for the pixels, and Figure 2.5(b) is the corresponding histogram with $b = 1$.

As mentioned above, like many other statistical features, histograms are typically used as features of a multimedia unit as a whole, such as an audio frame, a video frame, or an image. If we are interested in the features of the semantic objects captured in the multimedia data (e.g., just the horse in Figure 2.1 without caring about the background of other objects in the image), we need to first segment the objects in question from the multimedia data and then use the geometric features such as the moments that are variation-invariant for the objects, as histograms in general are variation-variant.

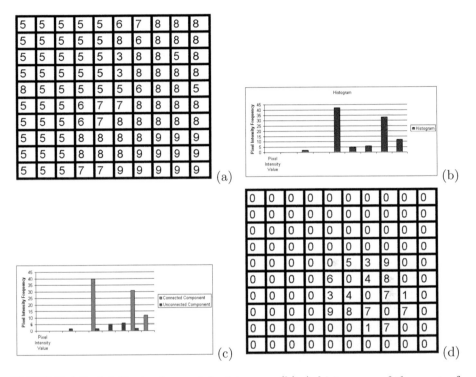

FIGURE 2.5: (a) Part of an original image; (b) A histogram of the part of the original image in (a) with the parameter $b = 1$; (c) A coherent vector of the part of the original image in (a) with the parameters $b = 1$ and $c = 5$; (d) The correlogram of the part of the original image in (a) with the parameters $b = 1$ and $k = 1$.

2.3.1.2   Coherent Vectors

Coherent vectors were first proposed in the early days of image retrieval in the mid-nineties [164]. They were used extensively in the early literature of image retrieval and were initially developed for color image retrieval. Since it is well-known that a histogram is not unique for representing a multimedia data unit, coherent vectors were proposed to improve this uniqueness.

Specifically, the idea of a coherent vector is to incorporate the spatial information into a histogram. Thus, a coherent vector is defined on top of a regular histogram, which is a vector. Given a regular histogram vector, data points in each component (called a *bucket*) of the histogram vector are further partitioned into two groups, one called *coherent* data points and the other called *incoherent* data points. A group of data points is defined as coherent if they are connected to form a connected component in the original domain of the multimedia data; otherwise, the data points are defined as incoherent. The specific implementation of the coherence definition is to set up a threshold c in advance such that a group of data points are coherent if their total count in the number of the data points that are connected exceeds c. Consequently, a coherent vector is a vector with each component as a pair of the number of the total coherent data points and the number of the total incoherent data points for the component.

Mathematically, if a regular histogram is represented as a regular vector

$$H = (h_1, ..., h_n)^T$$

then a coherent vector is represented as a vector of pairs

$$C = (\alpha_1 + \beta_1, ..., \alpha_n + \beta_n)^T$$

where α_i is the number of all the coherent data points in bucket i, β_i is the number of all the incoherent data points in bucket i, and $\alpha_i + \beta_i = h_i$, for all $i = 1, ..., n$. Figure 2.5(c) illustrates the coherent vector for the image shown in Figure 2.5(a) with parameters $b = 1, c = 5$.

2.3.1.3   Correlograms

Correlograms were another feature first proposed in the nineties in the image retrieval community [112]. Like coherent vectors, they were initially also developed for color image retrieval. The motivation for developing correlograms was also to further improve the representation uniqueness for this type of feature.

While coherent vectors incorporate the spatial information into the histogram features by labeling the data points in each bucket of the histogram into two groups — the coherent and the incoherent, through a connected component search — correlograms are a step further in incorporating the spatial information into the histogram features. Given a specific multimedia data unit of a specific modality of the data, a specific quantity $F(\boldsymbol{p})$ defined for a data point p in this unit, and a pre-defined distance value k between two data points in the unit, a correlogram of this unit is defined as a two-dimensional matrix C where each cell of the matrix $C(i,j)$ captures the frequency count of this unit for all the pairs of data points p_1 and p_2 such that $F(p_1) = i, F(p_2) = j$, and $d(p_1, p_2) = k$, where $d()$ is a distance function between two data points. Like a histogram, the dimensions of a correlogram C depend upon the granularity parameter b, with a larger b resulting in a "coarser" correlogram in lower dimensions and a smaller b resulting in a "finer" correlogram in higher dimensions. Figure 2.5(d) illustrates a correlogram of the original image in Figure 2.5(a) with parameters $b = 1, k = 1$, and the distance function $d()$ as the L_∞ metric, i.e., $d((x_1, y_1)^T, (x_2, y_2)^T) = \max(\|x_1 - x_2\|, \|y_1 - y_2\|)$.

2.3.1.4 Transformation Coefficient Features

Multimedia data are essentially all digital signals. As digital signals, all the different mathematical transformations may be applied to them to map them from their *original domains* to different domains typically called the *frequency domains*. Consequently, the coefficients of these transformations encode the statistical distributions of the multimedia data in their original domains as energy distributions in the frequency domains. Therefore, the coefficients of these transformations may also be used as features to represent the original multimedia data.

Since there are many mathematical transformations in the literature, different transformations result in different coefficient features. Here we review two important features that are often used in the literature, the Fourier transformation features and the wavelet transformation features. For simplicity purposes, we use one-dimensional multimedia data as an example. All the transformations may be applied to higher dimensional multimedia data.

Given a one-dimensional multimedia data sequence $f(x)$, its Fourier transformation is defined as follows [160]:

$$F(u) = \int_{-\infty}^{\infty} f(x)e^{-i2\pi xu}dx \qquad (2.5)$$

where $F(u)$ maps to the frequency domain and u is the variable in the frequency domain. Note that while typically $f(x)$ is a real function, $F(u)$ becomes a complex function based on this transformation. Due to this fact, $F(u)$ may be represented in the polar coordinate system as two real components, the amplitude $A(u)$ and the phase $\phi(u)$, i.e.,

$$F(u) = A(u)e^{i\phi(u)} \qquad (2.6)$$

In real-world applications, since $f(x)$ is always represented as a numerical series, the resulting coefficient functions $A(u)$ and $\phi(u)$ are also discrete series. In this case, the resulting transformation is actually called Discrete Fourier Transform (DFT), where $A(u)$ and $\phi(u)$ are both discrete sequences.

Consequently, either $A(U)$ or $\phi(u)$ alone or both of them may be used as the features for the original multimedia data $f(x)$. Given the fact that $f(x)$ can be completely recovered from Fourier Inverse Transformation [160]:

$$f(x) = \int_{-\infty}^{\infty} F(u)e^{i2\pi ux}du \qquad (2.7)$$

if we use both $A(u)$ and $\phi(u)$ as the whole series for the features of $f(x)$, these features are unique. Nevertheless, this would be the same as using the original data $f(x)$ themselves as the features and thus would have no benefits at all for the feature representation of the original data $f(x)$. Instead, we typically just truncate the series of $A(u)$ to the top few items as the features for $f(x)$, as typically the rest of the series items are very close to zero except for the first few items. The statistical interpretation of this practical truncation to generate the Fourier features is that those first few, non-zero items of $A(u)$ give a summarization of the global statistical distribution of the multimedia data, while the majority of the close-to-zero items of $A(u)$ indicate the many local details of the original data. In the literature, the first few items of $A(u)$ (corresponding to low u values) are called low-frequency components, whereas the rest of the items of $A(u)$ (corresponding to higher u values) are called high-frequency components. The reason why for many multimedia data the high-frequency components are always close to zero is that the high-frequency components represent the local changes, while the low-frequency components represent the global distributions; when we compare different multimedia data, the "thing" that makes them look different is the global distributions, whereas they often exhibit very similar local changes. Clearly, due to this truncation for representing Fourier features, the resulting features are no longer unique.

While Fourier coefficient features are good to capture the global information of multimedia data, in many applications it is important to pay attention to the local changes as well. In this case, wavelet transformation coefficient features are a good candidate for consideration.

Wavelet transformation is another very frequently used transformation. Given one-dimensional multimedia data $f(x)$, the classic wavelet transformation is defined as follows [93]:

$$W(a,b) = \frac{1}{\sqrt{a}} \int_{-\infty}^{\infty} f(x)\psi^*(\frac{x-b}{a})dx \qquad (2.8)$$

where a and b are the two variables to control the scaling and shift, respectively, of the transformation and $\psi()$ is called a mother function; $\psi^*()$ is its complex conjugate function. Like Fourier transformation, given the function $W(a,b)$ and a mother function $\psi()$, the original multimedia data $f(x)$ may be completely recovered from the wavelet inverse transformation [93]:

$$f(x) = C_\psi \int_{0}^{\infty} \int_{-\infty}^{\infty} \frac{1}{a^2\sqrt{a}} W(a,b)\psi(\frac{x-b}{a})dadb \qquad (2.9)$$

$$C_\psi = \int_0^\infty \frac{\psi^*(\nu)\psi\nu}{\nu} d\nu \tag{2.10}$$

From the definitions in Equations 2.8 and 2.9, it is clear that the wavelet transformation is more flexible than Fourier transformation for the following two reasons. First, the wavelet transformation allows different mother functions to be incorporated into the transform, while Fourier transformation can be considered as a special case of the wavelet transformation in the sense that the specific exponential function is used as the mother function. Second, the wavelet transformation involves two variables a and b controlling the scale and shift simultaneously for the transformation; since the scale reflects the space change and the shift reflects the time change, the wavelet transformation takes care of the time and space simultaneously to deliver a more powerful and flexible transformation. For these two reasons, the wavelet transformation is able to capture not only the global information but also the local changes as a good candidate for feature representation.

Like Fourier transformation, in real-world applications, the transformation $W(a, b)$ is always sampled into discrete series, with the variables a and b discretized into integers. Also, in real-world applications we are unable to keep the whole transformation series as the features because otherwise the features would be no better than having the original data as the representation. Consequently, like Fourier features, we always sample a few lower values of a and b, respectively, to obtain a few items of the wavelet transformation coefficients as the wavelet features. Due to this truncation, like Fourier features, the wavelet features are not unique, either.

2.3.2   Geometric Features

Unlike statistical features, geometric features are typically applied to segmented or identified objects within a multimedia data unit of a specific modality of the data. Consequently, a segmentation method must be first applied to a multimedia data unit to obtain such objects. Once such objects are obtained, geometric features are used to describe these objects. Due to this purpose, many geometric features are variation-invariant, offering the capability to preserve the same description while the objects are subject to different variations such as rotation changes, translation changes, and scale changes from one unit to another.

Depending upon how "completely" a specific geometric feature method is capable of describing an object in a multimedia data unit, some of the geometric features are unique, while others are not. Well-known geometric features include moments, inertia coefficients, and Fourier descriptors.

2.3.2.1   Moments

Moments also have a long history of being used as a type of geometric features for objects dating back to the early days of pattern recognition [67].

If we have the semantic objects segmented from the multimedia data, moments are a good candidate for the representation features, as they are variation-invariant; specifically, they are translation-invariant, rotation-invariant, and scale-invariant.

If an object is represented as a mathematical function $f(\boldsymbol{p})$ with respect to a coordinate vector $\boldsymbol{p}$ (e.g., if the object is a 1D signal, $\boldsymbol{p} = x$; if the object is a 2D signal, $\boldsymbol{p} = (x, y)^T$), the moments of the function $f(\boldsymbol{p})$ represent a series:

$$m_{\boldsymbol{q}} = \int_0^\infty \mathbf{1}^T \boldsymbol{p}^{\boldsymbol{q}} f(\boldsymbol{p}) d\boldsymbol{p} \qquad (2.11)$$

where $\boldsymbol{q}$ is an integer vector series where each component of the vector takes all non-negative integers independently, and $\boldsymbol{p}^{\boldsymbol{q}}$ are all the vectors where each component of $\boldsymbol{p}$ takes all the possible non-negative exponents of $\boldsymbol{q}$ independently and separately. For example, if $\boldsymbol{p}$ is a two-dimensional vector, i.e., $\boldsymbol{p} = (x, y)^T$, $\boldsymbol{p}^{\boldsymbol{q}} = \{(1, 1)^T, (1, y)^T, (1, y^2)^T, ..., (x, 1)^T, (x, y)^T, (x, y^2)^T, ..., (x^2, 1)^T, (x^2, y)^T, (x^2, y^2)^T, ...\}$. The vector $\boldsymbol{q}$ is called the *order* of the moments. The vector $\mathbf{1}$ is an all-unit-component vector, i.e., $\mathbf{1} = (1, 1, ..., 1)^T$ in the same dimension as that of $\boldsymbol{p}$.

There are two special moments, the zeroth-order moment $m_{\boldsymbol{0}}$ and the first order moment $m_{\boldsymbol{e}}$ where $\boldsymbol{e}$ is the full set of the basis vectors: $(1, 0, ..., 0)^T$, $(0, 1, ..., 0)^T, ...(0, 0, ..., 1)^T$. Clearly, while $m_{\boldsymbol{0}}$ is a single value, $m_{\boldsymbol{e}}$ is a vector of the same dimension as that of $\boldsymbol{p}$. We call $m_{\boldsymbol{0}}$ the *area* of the object and define the *center* of the object as

$$\bar{\boldsymbol{p}} = \frac{m_{\boldsymbol{e}}}{m_{\boldsymbol{0}}} \qquad (2.12)$$

Thus, we may change the original definition of the moments in Equation 2.11 to the *central moments* defined as follows:

$$\mu_{\boldsymbol{q}} = \int_0^\infty \mathbf{1}^T (\boldsymbol{p} - \bar{\boldsymbol{p}})^{\boldsymbol{q}} f(\boldsymbol{p}) d\boldsymbol{p} \qquad (2.13)$$

From the central moments, it is shown in the literature [109] that there is a set of invariants made from the central moments for translational, rotational, and scaling changes. It is also shown in the literature that, under certain conditions, if the whole (infinite) series of the moments is used, this feature is unique to represent a semantic object. Since in practice we can never use the whole infinite series, we always truncate to a limited few lower-order moments; this uniqueness property is compromised to a certain degree.

2.3.2.2   Fourier Descriptors

Like the moments, Fourier descriptors have also been used in a long history in the literature of pattern recognition [67]. Also like the moments, the Fourier descriptors can only be used when a semantic object is segmented from the multimedia data.

Given a segmented semantic object from the multimedia data, the idea of Fourier descriptors is to sample the contour of the segmented semantic object to form a sequence of the sampling points $p_1, p_2, \cdots, p_n$ on the contour, such as the point sequence in Figure 2.6(a), in a specific order (e.g., the counterclockwise order); then the sequence of the points may be transformed into the Fourier frequency domain by a Discrete Fourier Transform (DFT). This DFT sequence is called *Fourier descriptors* in the literature. Consequently, Fourier descriptors consist of two sequences, the amplitude $A(u)$ and the phase $\phi(u)$.

In the pattern recognition literature [67], it is shown that under a certain normalization, the amplitude sequence $A(u)$ alone exhibits an invariance under translation, rotation, scale, and the starting point of the original contour sequence variations. This is due to the fact that all these variations only contribute to a change in the phase sequence $\phi(u)$. Based on this fact, the amplitude sequence $A(u)$ may be used as a type of features for a segmented semantic object. On the other hand, due to this invariance fact, using $A(u)$ alone as the features, the features are not unique.

In order to accommodate extending the Fourier descriptors' invariance to the general affine transformation, in the literature, Zhang et al proposed to extend the existing Fourier descriptors to the *Area Based Fourier descriptors* [249]. The idea is to sample the object contour only for those "key" points instead of an arbitrary point or a uniform sampling such as that in Figure 2.6(a). A practical definition of the "key" points is those with a high curvature or with no curvature, such as those points $q_1, q_2, \cdots, q_n$ in Figure 2.6(b). Then we identify the center of the object O. Thus, instead of forming a sequence of the points along the contour in the Fourier descriptors, we now form a sequence of the areas along the contour where each area is determined by a pair of neighboring "key" points along the contour and the center of the object; i.e., each area A_i is defined by the points Oq_iq_{i+1}, as shown in Figure 2.6(b). It is shown [249] that under a certain normalization the amplitude sequence of the DFT of this area-based sequence is invariant under any affine transform. Therefore, this amplitude sequence can also be used as another type of feature for a segmented semantic object that is invariant under any affine transform. Due to this invariance, this type of feature is not unique, either.

2.3.2.3   Normalized Inertia Coefficients

Normalized inertia coefficients are actually a special case of the centralized moments representation. In the literature it was first proposed to be used as a feature to describe the shape of an object in an image [88]. Assuming a segmented object in a multimedia data unit has an area O, a complete description of the normalized inertia for an object in the multimedia data unit is a whole series with the *order* parameter q:

$$l(q) = \frac{\int_{\forall \boldsymbol{p} \in O} \|\boldsymbol{p} - \bar{\boldsymbol{p}}\|^{q/2} d\boldsymbol{p}}{\int_{\forall \boldsymbol{p} \in O} d\boldsymbol{p}} \tag{2.14}$$

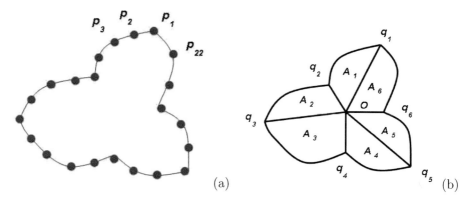

FIGURE 2.6: (a) The sequence of contour points sampled to form a Fourier descriptor. (b) The sequence of the areas to form an area-based Fourier descriptor.

where p is a point of the object O and $\bar{p}$ is the center point of the object.

Like other features, in practice, we always truncate the whole series to the first few lower q terms as the normalized inertia coefficients for the feature representation of the object O. Thus, this feature representation is not unique.

2.3.3   Meta Features

Meta features include the typical meta data to describe a multimedia data unit such as the scale of the unit, the number of objects in the unit, and the value range of the points in the unit.

2.4   Knowledge Representation

In order to effectively mine the multimedia data, it is important that not only an appropriate feature representation is used for the multimedia data but also that appropriate knowledge support is available in a multimedia database to facilitate the mining tasks. Like all other intelligent systems, a typical multimedia data mining system is often equipped with a knowledge support component to "guide" the mining tasks. Typical types of knowledge in the knowledge support component include the domain knowledge, the common sense knowledge, as well as the meta knowledge. Consequently, how to appropriately and effectively represent these types of knowledge in a multimedia data mining system has become a non-trivial research topic. On the other hand, like the general data mining activities, a typical multimedia data

FIGURE 2.7: (a) A natural scene image with mountains and blue sky. (b) An ideal labeling for the image.

mining task is to automatically discover the knowledge in a specific context. Thus, there is also a knowledge representation problem after the knowledge is discovered in the mining activity.

Knowledge representation has been one of the active core areas in artificial intelligence research, and many specific representation methods are proposed in the literature [29]. In the rest of this chapter, we first review several well-known knowledge representation methods in multimedia data mining applications and then demonstrate how to use these knowledge representation methods in different types of knowledge used in a multimedia data mining system.

2.4.1   Logic Representation

For human beings, a natural way to represent knowledge is through natural language. For example, in an imagery data mining system, if we want to restrict the domain to the natural scenery, we may want to have the following specific piece of knowledge: *All the blue areas indicate either sky or water.* This piece of knowledge would help generate the labels for the image in Figure 2.7(a) as shown in Figure 2.7(b), and that labeling further leads to knowledge discovery such as an answer to the question *What is considered the typical scene in the images of the database with a blue sky?*.

However, it is difficult, if not impossible, to make a computer system completely understand the natural language. An effective way to make the natural language understandable to a computer system is to use logic representation. A commonly used logic is called *predicate logic*, and the most popular predicate logic used in the literature is called *propositional first order logic*, typically abbreviated as FOL.

In FOL, all the variables are *set variables* in the sense that they can take any one of the values of the set defined for this variable. For example, a variable x may be defined as a real set $[0,1]$, which means that x may be any value in the domain of $[0,1]$; or the variable x may be defined as a *color set* variable, in which we may explicitly define $x = \{red, blue, white\}$ and x may take any of the three possible colors. All the functions in FOL are called *predicates*. All the predicates in FOL are *boolean predicates* in the sense that they can only return one of the two values: 0 for *false* and 1 for *true*. In addition, there are three operators defined for all the variables as well as the predicates. $\neg$ is a unitary operator applied to either a variable or a predicate resulting in a negation of the value of the operand; i.e., if the operand has a value 1, this operation has a value 0, and vice versa. $\wedge$ is a binary operator applied to either variables or predicates; it takes a multiplication between the values of the two operands; i.e., this operator returns 1 if and only if both values of the two operands are 1, and returns 0 otherwise. $\vee$ is another binary operator applied to either variables or predicates; it takes an addition between the two values of the two operands and, therefore, it returns 0 if and only if both operands have the values 0, and returns 1 otherwise. Finally, there are two quantifiers defined for variables *only*, but not for predicates. They are the *universal quantifier* $\forall$ and the *existential quantifier* $\exists$. $\forall$ means *for all the values* of the variable to which this quantifier applies. $\exists$ means *there exists at least one value* of the variable to which this quantifier applies.

Given this FOL, the natural language sentence *All the blue regions are either sky or water* may be represented as the following FOL statement:

$$\forall x (\text{blue}(x) \rightarrow \text{sky}(x) \vee \text{water}(x))$$

where blue(), sky(), and water() are the predicates, and $\rightarrow$ means "imply". On the other hand, the natural language sentence *Some blue regions are sky* may be represented as the following FOL statement:

$$\exists x (\text{blue}(x) \rightarrow \text{sky}(x))$$

The advantage of using FOL for knowledge representation in a multimedia data mining system is that FOL makes deductive reasoning very easy and powerful; the reasoning process is also very efficient due to the symbolic computation using the FOL statements. The disadvantage is when the knowledge base is dynamically and constantly updated from time to time, it is difficult to maintain the consistency of the FOL statements in the knowledge base for the multimedia data mining system.

2.4.2   Semantic Networks

Semantic networks are one very powerful knowledge representation tool used in today's artificial intelligence research and applications [199, 182]. They are proposed in the literature to use graphs to represent the concepts and

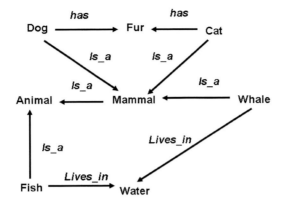

FIGURE 2.8: An example of a semantic network.

their relationships, in which the nodes in the graphs are the concepts and the edges in the graphs are the relationships. Historically, they were used to represent the English words as well as their relationships in natural language understanding research in artificial intelligence [182].

Figure 2.8 illustrates an example of a typical semantic network. In real world applications, the well-known WorldNet [1] is a good example of using semantic networks to represent the words and their relationships to serve as a lexical English word database. Note that the graph of a semantic network is a digraph, as the relationships are directional. Typical relationships include:

- *Meronymy:* A is part of B

- *Holonymy:* B is part of A

- *Hyponymy:* A is subordinate of B

- *Hypernymy:* A is superordinate of B

- *Synonymy:* A is the same as or similar to B

- *Antonymy:* A is the opposite of B

In multimedia data mining, semantic networks are used to represent the concepts, in particular, the spatial concepts, and their relationships. One example is the KMeD system developed by Hsu et al [107], which uses a hierarchical semantic network to represent the knowledge that is necessary for facilitating the reasoning, mining, and retrieval of the medical images in the database. Specifically, in the second layer of the hierarchy, called the semantic layer, objects and their relationships are identified and abstracted based on an entity-relationship model, and in the third layer a higher level of semantics called knowledge base abstraction is used to represent the domain expert knowledge using semantic networks to guide and improve the image mining and retrieval for the radiological image database.

Semantic networks represent loose associations between concepts. However, when they are used with added logical descriptions, semantic networks may have even more expressive power than FOL. Examples of these extensions of the semantic networks include the existential graphs [126] and the conceptual graphs [193].

2.4.3   Frames

Frames are another type of knowledge representation method used in a multimedia database for describing a specific type of object or a specific abstract concept. This knowledge representation method was initially due to Minsky [152]. A frame may have a name as well as other attributes which have values; these attributes are also called *slots*. Concepts related to each other may be described for one frame as a slot value of another frame. Minsky's original definition and description about frames is excerpted as follows [152].

> A *frame* is a data structure for representing a stereotyped situation, like being in a certain kind of living room or going to a child's birthday party. Attached to each frame are several kinds of information. Some of this information is about how to use the frame. Some is about what one can expect to happen next. Some is about what to do if these expectations are not confirmed.

> We can think of a frame as a network of nodes and relations. The "top levels" of a frame are fixed, and represent things that are always true about the supposed situation. The lower levels have many *terminals* — "slots" that must be filled by specific instances of data. Each terminal can specify conditions its argument must meet. (The assignments themselves are usually smaller "sub-frames.") Simple conditions are specified by markers that might require a terminal assignment to be a person, an object of sufficient value, or a pointer to a sub-frame of a certain type. More complex conditions can specify relations among the things assigned to several terminals.

For example, we may have the *house* frame to define and describe the related concepts below, where the frame *house* has slots of *style*, *color*, *door*, etc., and some slots are described by next levels of the frame. For example, the slot *door* is defined by the sub-frame *door-frame*. This can be further defined recursively. For example, the slot *garden* is described by the sub-frame *garden-frame*, which has a slot *pool* that is further defined by its sub-frame *pool-frame*; similarly, the slot *basement* is described by the sub-frame *basement-frame*, which has a lot *finished* that is further defined by its sub-frame *finished-basement-frame* if its value is not NULL, i.e., if the basement is finished. This whole *house* frame gives a network of nodes and relations to fully define what a typical two-story house looks like.

```
house:
  style: two story;
  color: white;
  door: door-frame;
  window: window-frame;
  room: room-frame;
  garden: garden-frame;
  garage: garage-frame;
  bathroom: bathroom-frame;
  basement: basement-frame;

door-frame:
  number-of-outside-door: 1;
  number-of-inside-door: 10;

garage-frame:
  size: 2;
  has-side-door: yes;

bathroom-frame:
  number-of-full: 3;
  number-of-half: 2;

window-frame:
  number-of-large: 3;
  number-of-small: 10;
  number-of-bay: 1;

room-frame:
  number-of-bedroom: 4;
  number-of-study-room: 1;
  number-of-living-room: 1;
  number-of-family-room: 1;
  number-of-workshop-room: 1;
  number-of-dining-room: 1;

garden-frame:
  fence: full;
  landscape: yes;
```

```
  lawn: yes;
  pool: pool-frame;
  number-of-trees: 3;
  area-in-acre: 0.2;

basement-frame:
  finished: finished-basement-frame;

finished-basement-frame:
  number-of-room: 2;
  number-of-bathroom: 1;
  number-of-storage-room: 1;
  walked-out: yes;

pool-frame:
  raised-or-ground: ground;
  area-in-square-feet: 150;
```

A typical example of using frames in knowledge representation in a multimedia database is the system proposed by Brink et al [30]. Specifically, frame abstractions are used to facilitate the encapsulation of file names, features, and relevant attributes of the image objects in the image database.

There is literature [100] arguing that frames have an equivalent expressive power in knowledge representation to that of logic. This is later demonstrated to be true to a certain extent [194].

2.4.4   Constraints

A constraint is a condition or a set of conditions that constrains the description of an object or an event. In the classic artificial intelligence research, many problems are described through constraints, and therefore *constraint satisfaction* is considered an effective approach to solving for those problems [182]. In multimedia data mining, typically there are three types of constraints:

- *Attribute Constraints*: Attribute constraints define the characteristic description of a multimedia object or a multimedia data item or an event. Examples include *The human face in this image is a male with red hair*; *The number of red color pixels in this image should be no more than 300*; and *The probability that there appears to be a moving target in the next five minutes of the surveillance video is about 0.3*.

- *Spatial Constraints*: Spatial constraints specify the spatial conditions that must hold true between multimedia data items or objects. Examples include *The person to the left of John Smith in this image is Harry*

Brown; *Located in the northwest corner of this map is the headquarters of ABC Inc.*; and *Pay attention to the top-right corner area in this video.*

- *Temporal Constraints*: Temporal constraints specify the temporal conditions that must hold true between multimedia data items or objects in an event. Examples include *John Smith shows up in the surveillance video before Harry Brown does*; *John Smith disappears in the surveillance video before Harry Brown does*; and *During all the time when John Smith shows up in the surveillance video, Harry Brown briefly shows up and disappears immediately.*

At the methodological level, constraints may be easily represented in terms of FOL sentences. If we are restricted only to the constraints of unitary or binary variables, which is the most typical scenario in multimedia data mining, we may also use a *constraint graph* to represent a set of constraints where each node represents a variable and each edge represents a binary constraint between the two variables; the unitary constraints are represented as attributes for the related variable nodes in the graph.

Given a set of constraints, finding a feasible solution with the presence of such a set of constraints is called *constraint satisfaction* in artificial intelligence research. Solutions to constraint satisfaction are part of the artificial intelligence search methods. Typical methods for constraint satisfaction include *generate and test, backtracking, forward checking*, and *constraint propagation*. *Generate and test* is a completely blind search method. The idea is to randomly generate the values for the whole set of the constrained variables and to test whether the values satisfy the whole set of constraints. Due to the "blindness" of the search, this method typically is very slow. *Backtracking* is another blind search method. Typically, backtracking is implemented with Depth-First-Search [182], in which, if any node along the search path violates a constraint, the search backtracks to the previous node to try an alternative path. *Forward checking* is yet another blind search method. In comparison with backtracking, this method further narrows down the search space by first lining up all the legitimate values for the constrained variables and then starting the search; during the search process, backtracking is used whenever a constraint is violated. *Constraint propagation* is also a blind search method but goes further than forward checking by propagating the constraints to obtaining the legitimate values of the constrained variables. Details of constraint satisfaction and the solutions may be found in [205].

A typical example of representing knowledge in constraints and using these constraints to facilitate the reasoning and mining in multimedia data is the *Show&Tell* system reported in [196]. Since this system is for mining aerial imagery data, only multimedia data of static imagery and text are used. Consequently, only characteristic constraints and spatial constraints are used in representing the knowledge; no temporal constraints are used. Figure 2.9 shows an example of the constraint satisfaction based spatial reasoning for

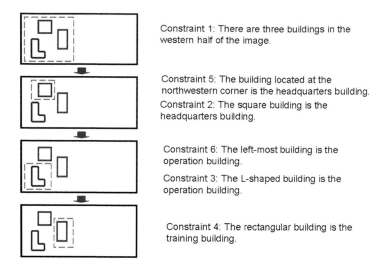

FIGURE 2.9: A hypothetical example to show how constraint satisfaction based reasoning helps mine the buildings from an aerial imagery database, where the dashed box indicates the current search focus.

mining the buildings in an aerial survey. Given the three buildings in the image and the following six constraints, where Constraints 2, 3, and 4 are characteristic constraints and Constraints 1, 5, and 6 are spatial constraints, Figure 2.9 demonstrates the constraint satisfaction based spatial reasoning for identifying all the three buildings. Specifically, Constraint 1 is first used to narrow down the focus of attention to the left of the image; then, based on Constraints 5 and 2, the square building located at the upper-left corner is identified as the headquarters building; then the focus of attention is moved to the leftmost, L-shaped building based on Constraints 6 and 3, and that building is immediately identified as the operation building; finally, using Constraint 4, the rectangular building is identified as the training building.

1. *Constraint 1*: There are three buildings in the western half of the image.

2. *Constraint 2*: The square building is the headquarters building.

3. *Constraint 3*: The L-shaped building is the operation building.

4. *Constraint 4*: The rectangular building is the training building.

5. *Constraint 5*: The building located at the northwestern corner is the headquarters building.

6. *Constraint 6*: The left-most building is the operation building.

2.4.5   Uncertainty Representation

While all the knowledge representation methods introduced above are for "certain" knowledge, in many real-world multimedia data mining problems, there are many occasions in which there is uncertainty in the knowledge. Consequently, we need to study how to represent the uncertainty in the knowledge.

In multimedia data mining, two commonly used approaches to representing uncertainties in knowledge are probability theory and fuzzy logic. In the following two sections, we review the two approaches and give examples in using these approaches in representing the uncertainties.

2.4.5.1   Using Probability Theory in Representing Uncertainties

Uncertainties in multimedia data mining may be represented in probability theory if we know a priori the probabilistic distributions of the variables in the data. In many scenarios we may not have this a priori knowledge. In this case, we must make assumptions for the a priori probabilistic distributions (e.g., we may assume a uniform prior probabilistic distribution for the variables). Given the prior probabilistic distributions, knowledge discovery is typically made possible through determining the posterior probabilities of the related variables in multimedia data. This is again typically achieved through Bayesian reasoning [182], and this reasoning may also be completed iteratively if there are latent variables involved (such as the Expectation and Maximization method [58]).

Chapter 6 focuses on a specific application example in using a probabilistic model to discover the hidden latent semantic concepts in an image database. Figure 2.10(a) shows a query image Im to be learned for the semantic concepts contained in the image, and Figure 2.10(b) shows the six image tokens $r_i(i = 1, \ldots, 6)$ learned in a method (see Chapter 6). Finally, Figure 2.10(c) shows the posterior probabilities $P(z_k|r_i, Im)$ for the concept *castle* z_k given each of the six tokens r_i in the image Im, as well as the posterior probability $P(z_k|Im)$ for the concept *castle* z_k given this image Im.

2.4.5.2   Using Fuzzy Logic to Represent Uncertainties

Uncertainties in multimedia data mining may be represented in fuzzy logic if we know in advance or if we may assume in advance the fuzzy membership function or the fuzzy resemblance function of the variables in a multimedia database. Fuzzy logic is derived from the fuzzy set theory [231] that deals with the typical approximate or fuzzy reasoning instead of the classic, precise reasoning in the predicate logic.

In fuzzy logic, variables are described in terms of the degree of truth typically represented in terms of a fuzzy membership function or a resemblance function that is mapped to an inclusive range of $[0, 1]$ [128, 260]. The degree of truth is often confused with a probability which is also mapped to an inclusive range of $[0, 1]$. However, they are conceptually different. Fuzzy

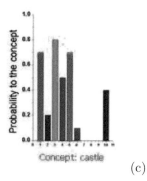

(a)   (b)   (c)

FIGURE 2.10: (a) An original query image. (b) The token image of the query image, where each token is represented as a unique color in the image. (c) The learned posterior probabilities for the concept *castle* in the query image.

truth represents membership in vaguely defined sets, whereas a probability of a certain event or condition states a deterministic likelihood. Below we use the example of using fuzzy logic to represent the fuzzy concept of a color histogram [242].

The color representation of an image pixel is coarse and imprecise if we simply extract the color feature of the pixel to index each region. Color is one of the most fundamental properties to discriminate images, so that we should take advantage of all available information in it. Considering the typical uncertainty stemming from color quantization and human perception, we develop a modified color histogram using the fuzzy technique [163, 211] to accommodate the uncertainty.

The fuzzy color histogram is defined as follows. We assume that each color is a fuzzy set, while the correlation among colors is modeled as a membership function of different fuzzy sets. A fuzzy set F on the feature space R^n is defined as a mapping $\mu_F : R^n \rightarrow [0, 1]$ with μ_F as the membership function. For any feature vector $f \in R^n$, the value of $\mu_F(f)$ is called the degree of membership of f to the fuzzy set F (or, in short, the degree of membership to F). A value closer to 1 for $\mu_F(f)$ means more representative the feature vector f is to the fuzzy set F.

An ideal fuzzy color model should have the resemblance inversely proportional to the inter-color distance. Based on this requirement, the most commonly used prototype membership functions include conic, trapezoidal, B-splines, exponential, Cauchy, and paired sigmoid functions [104]. We have tested the conic, trapezoidal, exponential, and Cauchy functions, respectively. In general, the performances[1] of the exponential and the Cauchy functions are better than those of the conic and trapezoidal functions. Considering the com-

---

[1]The performance means the average image retrieval accuracy given the same other settings.

putational complexity, we use the Cauchy function due to its computational simplicity. The Cauchy function, $C : R^n \rightarrow [0, 1]$, is defined as

$$C(\vec{x}) = \frac{1}{1 + (\frac{\|\vec{x} - \vec{v}\|}{d})^\alpha} \qquad (2.15)$$

where $\vec{v} \in R^n$, d and $\alpha \in R$, $d > 0$, $\alpha \geq 0$, $\vec{v}$ is the center location (point) of the fuzzy set, d represents the width of the function, and α determines the shape (or smoothness) of the function. Collectively, d and α describe the grade of fuzziness of the corresponding fuzzy feature.

Accordingly, the color resemblance in a region is defined as:

$$\mu_c(c') = \frac{1}{1 + (\frac{d(c,c')}{\sigma})^\alpha} \qquad (2.16)$$

where d is the Euclidean distance between color c and c' in the *Lab* color space, and σ is the average distance between colors

$$\sigma = \frac{2}{B(B-1)} \sum_{i=1}^{B-1} \sum_{k=i+1}^{B} d(c, c') \qquad (2.17)$$

where B is the number of the bins in the color partition. The average distance between colors is used to approximate the appropriate width of the fuzzy membership function. The experiments show that given the same other settings of the system, the average retrieval accuracy changes insignificantly when α is in the interval $[0.7, 1.5]$, but degrades rapidly outside the interval. We set $\alpha = 1$ in Equation 2.16 to simplify the computation.

This fuzzy color model enables us to enlarge the influence of a given color to its neighboring colors according to the uncertainty principle and the perceptual similarity. This means that each time a color c is found in the image, it influences all the quantized colors according to their resemblance to the color c. Numerically, this could be expressed as:

$$h_2(c) = \sum_{c' \in \mu} h_1(c')\mu_c(c') \qquad (2.18)$$

where μ is the color universe in the image and $h_1(c')$ is the standard normalized color histogram (see Section 2.3.1.1). Finally, the normalized fuzzy color histogram is computed as:

$$h(c) = \frac{h_2(c)}{\max_{c' \in \mu} h_2(c')} \qquad (2.19)$$

which falls in the range $[0,1]$.

Note that this fuzzy histogram computation is, in fact, a linear convolution between the standard color histogram and the fuzzy color model. This convolution expresses the histogram smoothing, provided that the color model is indeed a smoothing, low-pass filtering kernel. The use of the Cauchy function as the color model produces the smoothed histogram, which is a mean for the reduction of the quantization errors [122].

In the actual prototype implementation [242], the *Lab* color space is quantized into 96 buckets[2] by using the uniform quantization (*L* by 6, *a* by 4, and *b* by 4). To reduce the online computation, for each bin $\mu_c(c')$ is pre-computed and implemented as a lookup table.

2.5   Summary

In this chapter, we have introduced and reviewed the commonly used feature and knowledge representation methods in the context of multimedia data mining. Specifically, for feature representation, we have reviewed statistical features including histograms, coherent vectors, correlograms, and transformation coefficient features; geometric features including moments, Fourier descriptors, and normalized inertia coefficients; as well as meta features. For knowledge representation, we have reviewed methods using logic representation, semantic networks, frames, constraint representations, as well as uncertainty representations. These feature and knowledge representation methods have found extensive use and applications in all the multimedia data mining problems. Specifically, we have showcased their applications in solving different multimedia data mining problems in those chapters in Part III.

---

[2]Different numbers of buckets are tested, namely 96, 144, and 216 buckets; the difference on the final image retrieval accuracy is ignorable. For efficiency consideration, we use 96 buckets.

Chapter 3

Statistical Mining Theory and Techniques

3.1 Introduction

Multimedia data mining is an interdisciplinary research field in which generic data mining theory and techniques are applied to the multimedia data to facilitate multimedia-specific knowledge discovery tasks. In this chapter, commonly used and recently developed generic statistical learning theory, concepts, and techniques in recent multimedia data mining literature are introduced and their pros and cons are discussed. The principles and uniqueness of the applications of these statistical data learning and mining techniques to the multimedia domain are also provided in this chapter.

Data mining is defined as discovering hidden information in a data set. Like data mining in general, multimedia data mining involves many different algorithms to accomplish different tasks. All of these algorithms attempt to fit a model to the data. The algorithms examine the data and determine a model that is closest to the characteristics of the data being examined. Typical data mining algorithms can be characterized as consisting of three components:

- *Model*: The purpose of the algorithm is to fit a model to the data.

- *Preference*: Some criteria must be used to select one model over another.

- *Search*: All the algorithms require searching the data.

The model in data mining can be either predictive or descriptive in nature. A *predictive model* makes a prediction about values of data using known results found from different data sources. A *descriptive model* identifies patterns or relationships in data. Unlike the predictive model, a descriptive model serves as a way to explore the properties of the data examined, not to predict new properties.

There are many different statistical methods used to accommodate different multimedia data mining tasks. These methods not only require specific types of data structures, but also imply certain types of algorithmic approaches. The statistical learning theory and techniques introduced in this chapter are the ones that are commonly used in practice and/or recently developed in the literature to perform specific multimedia data mining tasks as exemplified in the subsequent chapters of the book. Specifically, in the multimedia data mining context, the classification and regression tasks are especially pervasive, and the data-driven statistical machine learning theory and techniques are particularly important. Two major paradigms of statistical learning models that are extensively used in the recent multimedia data mining literature are studied and introduced in this chapter: the *generative models* and the *discriminative models*. In the generative models, we mainly focus on the Bayesian learning, ranging from the classic Naive Bayes Learning, to the Belief Networks, to the most recently developed graphical models including Latent Dirichlet Allocation, Probabilistic Latent Semantic Analysis, and Hierarchical Dirichlet Process. In the discriminative models, we focus on the Support Vector Machines, as well as its recent development in the context of multimedia data mining on maximum margin learning with structured output space, and the Boosting theory for combining a series of weak classifiers into a stronger one. Considering the typical special application requirements in multimedia data mining where it is common that we encounter ambiguities and/or scarce training samples, we also introduce two recently developed learning paradigms: multiple instance learning and semi-supervised learning, with their applications in multimedia data mining. The former addresses the training scenario when ambiguities are present, while the latter addresses the training scenario when there are only a few training samples available. Both these scenarios are very common in multimedia data mining and, therefore, it is important to include these two learning paradigms into this chapter.

The remainder of this chapter is organized as follows. Section 3.2 introduces Bayesian learning. A well-studied statistical analysis technique, Probabilistic Latent Semantic Analysis, is introduced in Section 3.3. Section 3.4 introduces another related statistical analysis technique, Latent Dirichlet Allocation (LDA), and Section 3.5 introduces the most recent extension of LDA to a hierarchical learning model called Hierarchical Dirichlet Process (HDP). Section 3.6 briefly reviews the recent literature in multimedia data mining using these generative latent topic discovery techniques. Afterwards, an important, and probably the most important, discriminative learning model, Support Vector Machines, is introduced in Section 3.7. Section 3.8 introduces the recently developed maximum margin learning theory in the structured output space with its application in multimedia data mining. Section 3.9 introduces the boosting theory to combine multiple weak learners to build a strong learner. Section 3.10 introduces the recently developed multiple instance learning theory and its applications in multimedia data mining. Section 3.11 introduces another recently developed learning theory with extensive multimedia data mining applications called semi-supervised learning. Finally, this chapter is summarized in Section 3.12.

3.2 Bayesian Learning

Bayesian reasoning provides a probabilistic approach to inference. It is based on the assumption that the quantities of interest are governed by probability distribution and that an optimal decision can be made by reasoning about these probabilities together with observed data. A basic familiarity with Bayesian methods is important to understand and characterize the operation of many algorithms in machine learning. Features of Bayesian learning methods include:

- Each observed training example can incrementally decrease or increase the estimated probability that a hypothesis is correct. This provides a more flexible approach to learning than algorithms that completely eliminate a hypothesis if it is found to be inconsistent with any single example.

- Prior knowledge can be combined with the observed data to determine the final probability of a hypothesis. In Bayesian learning, prior knowledge is provided by asserting (1) a prior probability for each candidate hypothesis, and (2) a probability distribution over the observed data for each possible hypothesis.

- Bayesian methods can accommodate hypotheses that make probabilistic predictions (e.g., the hypothesis such as "this email has a 95% probability of being spam").

- New instances can be classified by combining the predictions of multiple hypotheses, weighted by their probabilities.

- Even in cases where Bayesian methods prove computationally intractable, they can provide a standard of optimal decision making against which other practical methods can be measured.

3.2.1 Bayes Theorem

In multimedia data mining we are often interested in determining the best hypothesis from a space H, given the observed training data D. One way to specify what we mean by the *best* hypothesis is to say that we demand the most probable hypothesis, given the data D plus any initial knowledge about the prior probabilities of the various hypotheses in H. Bayes theorem provides a direct method for calculating such probabilities. More precisely, Bayes theorem provides a way to calculate the probability of a hypothesis based on its prior probability, the probabilities of observing various data given the hypothesis, and the observed data themselves.

First, let us introduce the notations. We shall write $P(h)$ to denote the initial probability that hypothesis h holds true, before we have observed the training data. $P(h)$ is often called the *prior probability* of h and may reflect any background knowledge we have about the chance that h is a correct hypothesis. If we have no such prior knowledge, then we might simply assign the same prior probability to each candidate hypothesis. Similarly, we will write $P(D)$ to denote the prior probability that training data set D is observed (i.e., the probability of D given no knowledge about which the hypothesis holds true). Next we write $P(D|h)$ to denote the probability of observing data D given a world in which hypothesis h holds true. More generally, we write $P(x|y)$ to denote the probability of x given y. In machine learning problems we are interested in the probability $P(h|D)$ that h holds true given the observed training data D. $P(h|D)$ is called the *posterior probability* of h, because it reflects our confidence that h holds true after we have seen the training data D. Note that the posterior probability $P(h|D)$ reflects the influence of the training data D, in contrast to the prior probability $P(h)$, which is independent of D.

Bayes theorem is the cornerstone of Bayesian learning methods because it provides a way to compute the posterior probability $P(h|D)$ from the prior probability $P(h)$, together with $P(D)$ and $P(D|h)$. Bayes Theorem states:

THEOREM 3.1

$$P(h|D) = \frac{P(D|h)P(h)}{P(D)} \tag{3.1}$$

As one might intuitively expect, $P(h|D)$ increases with $P(h)$ and with $P(D|h)$, according to Bayes theorem. It is also reasonable to see that $P(h|D)$ decreases as $P(D)$ increases, because the more probably D is observed independent of h, the less evidence D provides in support of h.

In many classification scenarios, a learner considers a set of candidate hypotheses H and is interested in finding the most probable hypothesis $h \in H$ given the observed data D (or at least one of the maximally probable hypotheses if there are several). Any such maximally probable hypothesis is called a *maximum a posteriori* (MAP) hypothesis. We can determine the MAP hypotheses by using Bayes theorem to compute the posterior probability of each candidate hypothesis. More precisely, we say that h_{MAP} is a MAP hypothesis provided

$$h_{MAP} = \arg\max_{h \in H} P(h|D)$$
$$= \arg\max_{h \in H} \frac{P(D|h)P(h)}{P(D)}$$
$$= \arg\max_{h \in H} P(D|h)P(h) \tag{3.2}$$

Notice that in the final step above we have dropped the term $P(D)$ because it is a constant independent of h.

Sometimes, we assume that every hypothesis in H is equally probable a priori ($P(h_i) = P(h_j)$ for all h_i and h_j in H). In this case we can further simplify Equation 3.2 and need only consider the term $P(D|h)$ to find the most probable hypothesis. $P(D|h)$ is often called the *likelihood* of the data D given h, and any hypothesis that maximizes $P(D|h)$ is called a *maximum likelihood* (ML) hypothesis, h_{ML}.

$$h_{ML} \equiv \arg\max_{h \in H} P(D|h) \tag{3.3}$$

3.2.2 Bayes Optimal Classifier

The previous section introduces Bayes theorem by considering the question "What is the most probable hypothesis given the training data?" In fact, the question that is often of most significance is the closely related question "What is the most probable classification of the new instance given the training data?" Although it may seem that this second question can be answered by simply applying the MAP hypothesis to the new instance, in fact, it is possible to even do things better.

To develop an intuition, consider a hypothesis space containing three hypotheses, h_1, h_2, and h_3. Suppose that the posterior probabilities of these hypotheses given the training data are 0.4, 0.3, and 0.3, respectively. Thus, h_1 is the MAP hypothesis. Suppose a new instance x is encountered, which is classified positive by h_1 but negative by h_2 and h_3. Taking all hypotheses into account, the probability that x is positive is 0.4 (the probability associated with h_1), and the probability that it is negative is therefore 0.6. The most probable classification (negative) in this case is different from the classification generated by the MAP hypothesis.

In general, the most probable classification of a new instance is obtained by combining the predictions of all hypotheses, weighted by their posterior probabilities. If the possible classification of the new example can take on any value v_j from a set V, then the probability $P(v_j|D)$ that the correct classification for the new instance is v_j is just

$$P(v_j|D) = \sum_{h_i \in H} P(v_j|h_i) P(h_i|D)$$

The optimal classification of the new instance is the value v_j, for which $P(v_j|D)$ is maximum. Consequently, we have the Bayes optimal classification:

$$\arg\max_{v_j \in V} \sum_{h_i \in H} P(v_j|h_i) P(h_i|D) \tag{3.4}$$

Any system that classifies new instances according to Equation 3.4 is called a *Bayes optimal classifier*, or Bayes optimal learner. No other classification method using the same hypothesis space and the same prior knowledge can outperform this method on average. This method maximizes the probability that the new instance is classified correctly, given the available data, hypothesis space, and prior probabilities over the hypotheses.

Note that one interesting property of the Bayes optimal classifier is that the predictions it makes can correspond to a hypothesis not contained in H. Imagine using Equation 3.4 to classify every instance in X. The labeling of instances defined in this way need not correspond to the instance labeling of any single hypothesis h from H. One way to view this situation is to think of the Bayes optimal classifier as effectively considering a hypothesis space H' different from the space of hypotheses H to which Bayes theorem is being applied. In particular, H' effectively includes hypotheses that perform comparisons between linear combinations of predictions from multiple hypotheses in H.

3.2.3   Gibbs Algorithm

Although the Bayes optimal classifier obtains the best performance that can be achieved from the given training data, it may also be quite costly to apply. The expense is due to the fact that it computes the posterior probability for every hypothesis in H and then combines the predictions of each hypothesis to classify each new instance.

An alternative, less optimal method is the Gibbs algorithm [161], defined as follows:

1. Choose a hypothesis h from H at random, according to the posterior probability distribution over H.

2. Use h to predict the classification of the next instance x.

Given a new instance to classify, the Gibbs algorithm simply applies a hypothesis drawn at random according to the current posterior probability distribution. Surprisingly, it can be shown that under certain conditions the expected misclassification error for the Gibbs algorithm is at most twice the expected error of the Bayes optimal classifier. More precisely, the expected value is taken over target concepts drawn at random according to the prior probability distribution assumed by the learner. Under this condition, the expected value of the error of the Gibbs algorithm is at worst twice the expected value of the error of the Bayes optimal classifier.

3.2.4   Naive Bayes Classifier

One highly practical Bayesian learning method is the naive Bayes learner, often called the *naive Bayes classifier*. In certain domains its performance has been shown to be comparable to those of neural network and decision tree learning.

The naive Bayes classifier applies to learning tasks where each instance x is described by a conjunction of attribute values and where the target function $f(x)$ can take on any value from a finite set V. A set of training examples of the target function is provided, and a new instance is presented, described by the tuple of attribute values $(a_1, a_2, ..., a_n)$. The learner is asked to predict the target value, or classification, for this new instance.

The Bayesian approach to classifying the new instance is to assign the most probable target value, v_{MAP}, given the attribute values $(a_1, a_2, ..., a_n)$ that describe the instance.

$$v_{MAP} = \arg\max_{v_j \in V} P(v_j | a_1, a_2, ..., a_n)$$

We can use Bayes theorem to rewrite this expression as

$$v_{MAP} = \arg\max_{v_j \in V} \frac{P(a_1, a_2, ..., a_n | v_j) P(v_j)}{P(a_1, a_2, ..., a_n)}$$
$$= \arg\max_{v_j \in V} P(a_1, a_2, ..., a_n | v_j) P(v_j) \qquad (3.5)$$

Now we can attempt to estimate the two terms in Equation 3.5 based on the training data. It is easy to estimate each of the $P(v_j)$ simply by counting the frequency in which each target value v_j occurs in the training data. However, estimating the different $P(a_1, a_2, ..., a_n | v_j)$ terms in this fashion is not feasible unless we have a very large set of training data. The problem is that the number of these terms is equal to the number of possible instances times the number of possible target values. Therefore, we need to see every instance in the instance space many times in order to obtain reliable estimates.

The naive Bayes classifier is based on the simplifying assumption that the attribute values are conditionally independent given the target value. In other words, the assumption is that given the target value of the instance, the probability of observing the conjunction $a_1, a_2, ...a_n$ is just the product of the probabilities for the individual attributes: $P(a_1, a_2, ..., a_n | v_j) = \prod_i P(a_i | v_j)$. Substituting this into Equation 3.5, we have the approach called the *naive Bayes classifier*:

$$v_{NB} = \arg\max_{v_j \in V} P(v_j) \prod_i P(a_i | v_j) \qquad (3.6)$$

where v_{NB} denotes the target value output by the naive Bayes classifier. Notice that in a naive Bayes classifier the number of distinct $P(a_i | v_j)$ terms that must be estimated from the training data is just the number of distinct attribute values times the number of distinct target values — a much smaller number than if we were to estimate the $P(a_1, a_2, ..., a_n | v_j)$ terms as first contemplated.

To summarize, the naive Bayes learning method involves a learning step in which the various $P(v_j)$ and $P(a_i | v_j)$ terms are estimated, based on their frequencies over the training data. The set of these estimates corresponds to the learned hypothesis. This hypothesis is then used to classify each new instance by applying the rule in Equation 3.6. Whenever the naive Bayes assumption of conditional independence is satisfied, this naive Bayes classification v_{NB} is identical to the MAP classification.

One interesting difference between the naive Bayes learning method and other learning methods is that there is no explicit search through the space of possible hypotheses (in this case, the space of possible hypotheses is the space of possible values that can be assigned to the various $P(v_j)$ and $P(a_i|v_j)$ terms). Instead, the hypothesis is formed without searching, simply by counting the frequency of various data combinations within the training examples.

3.2.5   Bayesian Belief Networks

As discussed in the previous two sections, the naive Bayes classifier makes significant use of the assumption that the values of the attributes $a_1, a_2, ..., a_n$ are conditionally independent given the target value v. This assumption dramatically reduces the complexity of learning the target function. When it is met, the naive Bayes classifier outputs the optimal Bayes classification. However, in many cases this conditional independence assumption is clearly overly restrictive.

A Bayesian belief network describes the probability distribution governing a set of variables by specifying a set of conditional independence assumptions along with a set of conditional probabilities. In contrast to the naive Bayes classifier, which assumes that *all* the variables are conditionally independent given the value of the target variable, Bayesian belief networks allow stating conditional independence assumptions that apply to *subsets* of the variables. Thus, Bayesian belief networks provide an intermediate approach that is less constraining than the global assumption of conditional independence made by the naive Bayes classifier, but more tractable than avoiding conditional independence assumptions altogether. Bayesian belief networks are an active focus of current research, and a variety of algorithms have been proposed for learning them and for using them for inference. In this section we introduce the key concepts and the representation of Bayesian belief networks.

In general, a Bayesian belief network describes the probability distribution over a set of variables. Consider an arbitrary set of random variables $Y_1, ..., Y_n$, where each variable Y_i can take on the set of possible values $V(Y_i)$. We define the *joint space* of the set of variables Y to be the cross product $V(Y_1) \times V(Y_2) \times ...V(Y_n)$. In other words, each item in the joint space corresponds to one of the possible assignments of values to the tuple of variables $(Y_1, ..., Y_n)$. A Bayesian belief network describes the joint probability distribution for a set of variables.

Let X, Y, and Z be three discrete-value random variables. We say that X is conditionally independent of Y given Z if the probability distribution governing X is independent of the value of Y given a value for Z; that is, if

$$(\forall x_i, y_j, z_k) P(X = x_i | Y = y_j, Z = z_k) = P(X = x_i | Z = z_k)$$

where $x_i \in V(X), y_j \in V(Y), z_k \in V(Z)$. We commonly write the above expression in the abbreviated form $P(X|Y, Z) = P(X|Z)$. This definition of conditional independence can be extended to sets of variables as well. We say that the set of variables $X_1...X_l$ is conditionally independent of the set of variables $Y_1...Y_m$ given the set of variables $Z_1...Z_n$ if

$$P(X_1...X_l | Y_1...Y_m, Z_1...Z_n) = P(X_1...X_l | Z_1...Z_n)$$

Note the correspondence between this definition and our use of the conditional independence in the definition of the naive Bayes classifier. The naive Bayes classifier assumes that the instance attribute A_1 is conditionally independent of instance attribute A_2 given the target value V. This allows the naive Bayes classifier to compute $P(A_1, A_2 | V)$ in Equation 3.6 as follows:

$$\begin{aligned} P(A_1, A_2 | V) &= P(A_1 | A_2, V) P(A_2 | V) \\ &= P(A_1 | V) P(A_2 | V) \end{aligned} \tag{3.7}$$

A Bayesian belief network (Bayesian network for short) represents the joint probability distribution for a set of variables. In general, a Bayesian network represents the joint probability distribution by specifying a set of conditional independence assumptions (represented by a directed acyclic graph), together with sets of local conditional probabilities. Each variable in the joint space is represented by a node in the Bayesian network. For each variable two types of information are specified. First, the network arcs represent the assertion that the variable is conditionally independent of its nondescendants in the network given its immediate predecessors in the network. We say X is a *descendant* of Y if there is a directed path from Y to X. Second, a conditional probability table is given for each variable, describing the probability distribution for that variable given the values of its immediate predecessors. The joint probability for any desired assignment of values $(y_1, ..., y_n)$ to the tuple of network variables $(Y_1...Y_n)$ can be computed by the formula

$$P(y_1, ..., y_n) = \prod_{i=1}^{n} P(y_i | Parents(Y_i))$$

where $Parents(Y_i)$ denotes the set of immediate predecessors of Y_i in the network. Note that the values of $P(y_i | Parents(Y_i))$ are precisely the values stored in the conditional probability table associated with node Y_i. Figure 3.1 shows an example of a Bayesian network. Associated with each node is a set of conditional probability distributions. For example, the "Alarm" node might have the probability distribution shown in Table 3.1.

We might wish to use a Bayesian network to infer the value of a target variable given the observed values of the other variables. Of course, given the fact

Multimedia Data Mining

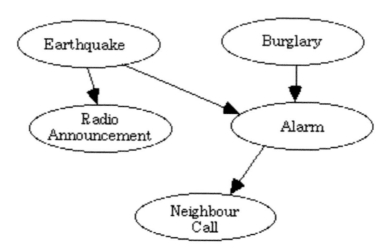

FIGURE 3.1: Example of a Bayesian network.

Table 3.1: Associated conditional probabilities with the node "Alarm" in Figure 3.1.

E	B	$P(A|E, B)$	$P(\neg A|E, B)$
E	B	0.90	0.10
E	$\neg B$	0.20	0.80
$\neg E$	B	0.90	0.10
$\neg E$	$\neg B$	0.01	0.99

that we are dealing with random variables, it is not in general correct to assign the target variable a single determined value. What we really wish to refer to is the probability distribution for the target variable, which specifies the probability that it will take on each of its possible values given the observed values of the other variables. This inference step can be straightforward if the values for all of the other variables in the network are known exactly. In the more general case, we may wish to infer the probability distribution for some variables given observed values for only a subset of the other variables. Generally speaking, a Bayesian network can be used to compute the probability distribution for any subset of network variables given the values or distributions for any subset of the remaining variables.

Exact inference of probabilities in general for an arbitrary Bayesian network is known to be NP-hard [51]. Numerous methods have been proposed for probabilistic inference in Bayesian networks, including exact inference methods and approximate inference methods that sacrifice precision to gain efficiency. For example, Monte Carlo methods provide approximate solutions by randomly sampling the distributions of the unobserved variables [170]. In theory, even approximate inference of probabilities in Bayesian networks can be NP-hard [54]. Fortunately, in practice approximate methods have been shown to be useful in many cases.

In the case where the network structure is given in advance and the variables are fully observable in the training examples, learning the conditional probability tables is straightforward. We simply estimate the conditional probability table entries just as we would for a naive Bayes classifier. In the case where the network structure is given but only the values of some of the variables are observable in the training data, the learning problem is more difficult. This problem is somewhat analogous to learning the weights for the hidden units in an artificial neural network, where the input and output node values are given but the hidden unit values are left unspecified by the training examples. Similar gradient ascent procedures that learn the entries in the conditional probability tables have been proposed, such as [182]. The gradient ascent procedures search through a space of hypotheses that corresponds to the set of all possible entries for the conditional probability tables. The objective function that is maximized during gradient ascent is the probability $P(D|h)$ of the observed training data D given the hypothesis h. By definition, this corresponds to searching for the maximum likelihood hypothesis for the table entries.

Learning Bayesian networks when the network structure is not known in advance is also difficult. Cooper and Herskovits [52] present a Bayesian scoring metric for choosing among alternative networks. They also present a heuristic search algorithm for learning network structure when the data are fully observable. The algorithm performs a greedy search that trades off network complexity for accuracy over the training data. Constraint-based approaches to learning Bayesian network structure have also been developed [195]. These approaches infer independence and dependence relationships from the data, and then use these relationships to construct Bayesian networks.

3.3 Probabilistic Latent Semantic Analysis

One of the fundamental problems in mining from textual and multimedia data is to learn the meaning and usage of data objects in a data-driven fashion, e.g., from given images or video keyframes, possibly without further domain prior knowledge. The main challenge a machine learning system has to address is rooted in the distinction between the lexical level of "what actually has been shown" and the semantic level of what "what was intended" or "what was referred to" in a multimedia data unit. The resulting problem is two-fold: (i) polysemy, i.e., a unit may have multiple senses and multiple types of usage in different contexts, and (ii) synonymy and semantically related units, i.e., different units may have a similar meaning; they may, at least in certain contexts, denote the same concept or refer to the same topic.

Latent semantic analysis (LSA) [56] is a well-known technique which partially addresses these questions. The key idea is to map high-dimensional count vectors, such as the ones arising in vector space representations of multimedia units, to a lower-dimensional representation in a so-called *latent semantic space*. As the name suggests, the goal of LSA is to find a data mapping which provides information well beyond the lexical level and reveals semantic relations between the entities of interest. Due to its generality, LSA has proven to be a valuable analysis tool with a wide range of applications. Despite its success, there are a number of downsides of LSA. First of all, the methodological foundation remains to a large extent unsatisfactory and incomplete. The original motivation for LSA stems from linear algebra and is based on L_2-optimal approximation of matrices of unit counts based on the *Singular Value Decomposition* (SVD) method. While SVD by itself is a well-understood and principled method, its application to count data in LSA remains somewhat *ad hoc*. From a statistical point of view, the utilization of the L_2-norm approximation principle is reminiscent of a Gaussian noise assumption which is hard to justify in the context of count variables. At a deeper, conceptual level the representation obtained by LSA is unable to handle polysemy. For example, it is easy to show that in LSA the coordinates of a word in a latent space can be written as a linear superposition of the coordinates of the documents that contain the word. The superposition principle, however, is unable to explicitly capture multiple senses of a word (i.e., a unit), and it does not take into account that every unit occurrence is typically intended to refer to one meaning at a time.

Probabilistic Latent Semantic Analysis (pLSA), also known as *Probabilistic Latent Semantic Indexing* (pLSI) in the literature, stems from a statistical view of LSA. In contrast to the standard LSA, pLSA defines a proper generative data model. This has several advantages as follows. At the most general level it implies that standard techniques from statistics can be applied for model fitting, model selection, and complexity control. For example, one can assess the quality of the pLSA model by measuring its predictive performance, e.g., with the help of cross-validation. At the more specific level, pLSA associates a latent context variable with each unit occurrence, which explicitly accounts for polysemy.

3.3.1 Latent Semantic Analysis

LSA can be applied to any type of count data over a discrete dyadic domain, known as the *two-mode data*. However, since the most prominent application of LSA is in the analysis and retrieval of text documents, we focus on this setting for the introduction purpose in this section. Suppose that we are given a collection of text documents $D = d_1, ..., d_N$ with terms from a vocabulary $W = w_1, ..., w_M$. By ignoring the sequential order in which words occur in a document, one may summarize the data in a rectangular $N \times M$ *co-occurrence table* of counts $N = (n(d_i, w_j))_{ij}$, where $n(d_i, w_j)$ denotes the number of the times the term w_j has occurred in document d_i. In this particular case, N is also called the term-document matrix and the rows/columns of N are referred to as document/term vectors, respectively. The key assumption is that the simplified "bag-of-words" or vector-space representation of the documents will in many cases preserve most of the relevant information, e.g., for tasks such as text retrieval based on keywords.

The co-occurrence table representation immediately reveals the problem of *data sparseness*, also known as the *zero-frequency* problem. A typical term-document matrix derived from short articles, text summaries, or abstracts may only have a small fraction of non-zero entries, which reflects the fact that only very few of the words in the vocabulary are actually used in any single document. This has problems, for example, in the applications that are based on matching queries against documents or evaluating similarities between documents by comparing common terms. The likelihood of finding many common terms even in closely related articles may be small, just because they might not use *exactly* the same terms. For example, most of the matching functions used in this context are based on similarity functions that rely on inner products between pairs of document vectors. The encountered problems are then two-fold: On the one hand, one has to account for synonyms in order not to underestimate the true similarity between the documents. On the other hand, one has to deal with polysems to avoid overestimating the true similarity between the documents by counting common terms that are used in different meanings. Both problems may lead to inappropriate lexical matching scores which may not reflect the "true" similarity hidden in the semantics of the words.

As mentioned previously, the key idea of LSA is to map documents — and, by symmetry, terms — to a vector space in a reduced dimensionality, the *latent semantic space*, which in a typical application in document indexing is chosen to have an order of about 100–300 dimensions. The mapping of the given document/term vectors to their latent space representatives is restricted to be linear and is based on decomposition of the co-occurrence matrix N by SVD. One thus starts with the standard SVD given by

$$N = USV^T \tag{3.8}$$

where U and V are matrices with orthonormal columns $U^T U = V^T V = I$ and the diagonal matrix S contains the singular values of N. The LSA approximation of N is computed by thresholding all but the largest K singular values in S to zero ($= \tilde{S}$), which is rank K optimal in the sense of the L_2-matrix or Frobenius norm, as is well-known from linear algebra; i.e., we have the approximation

$$\tilde{N} = U\tilde{S}V^T \approx USV^T = N \tag{3.9}$$

Note that if we want to compute the document-to-document inner products based on Equation 3.9, we would obtain $\tilde{N}\tilde{N}^T = U\tilde{S}^2 U^T$, and hence one might think of the rows of $U\tilde{S}$ as defining coordinates for documents in the latent space. While the original high-dimensional vectors are sparse, the corresponding low-dimensional latent vectors are typically not sparse. This implies that it is possible to compute meaningful association values between pairs of documents, even if the documents do not have any terms in common. The hope is that terms having a common meaning are roughly mapped to the same direction in the latent space.

3.3.2 Probabilistic Extension to Latent Semantic Analysis

The starting point for *probabilistic latent semantic analysis* [101] is a statistical model which has been called the *apsect model*. In the statistical literature similar models have been discussed for the analysis of contingency tables. Another closely related technique called non-negative matrix factorization [135] has also been proposed. The aspect model is a latent variable model for co-occurrence data which associates an unobserved class variable $z_k \in \{z_1, ..., z_K\}$ with each observation, an observation being the occurrence of a word in a particular document. The following probabilities are introduced in pLSA: $P(d_i)$ is used to denote the probability that a word occurrence is observed in a particular document d_i; $P(w_j|z_k)$ denotes the class-conditional probability of a specific word conditioned on the unobserved class variable z_k; and, finally, $P(z_k|d_i)$ denotes a document-specific probability distribution over the latent variable space. Using these definitions, one may define a generative model for words/document co-occurrences by the scheme [161] defined as follows:

1. select a document d_i with probability $P(d_i)$;

2. pick a latent class z_k with probability $P(z_k|d_i)$;

3. generate a word w_j with probability $P(w_j|z_k)$.

As a result, one obtains an observation pair (d_i, w_j), while the latent class variable z_k is discarded. Translating the data generation process into a joint probability model results in the expression:

$$P(d_i, w_j) = P(d_i)P(w_j|d_i) \qquad (3.10)$$

$$P(w_j|d_i) = \sum_{k=1}^{K} P(w_j|z_k)P(z_k|d_i) \qquad (3.11)$$

Essentially, to obtain Equation 3.11 one has to sum over the possible choices of z_k by which an observation could have been generated. Like virtually all the statistical latent variable models, the aspect model introduces a conditional independence assumption, namely that d_i and w_j are independent, conditioned on the state of the associated latent variable. A very intuitive interpretation for the aspect model can be obtained by a close examination of the conditional distribution $P(w_j|d_i)$, which is seen to be a convex combination of the k class-conditionals or *aspects* $P(w_j|z_k)$. Loosely speaking, the modeling goal is to identify conditional probability mass functions $P(w_j|z_k)$ such that the document-specific word distributions are as faithfully as possible approximated by the convex combinations of these aspects. More formally, one can use a maximum likelihood formulation of the learning problem; i.e., one has to maximize

$$\mathcal{L} = \sum_{i=1}^{N}\sum_{j=1}^{M} n(d_i, w_j) \log P(d_i, w_j)$$

$$= \sum_{i=1}^{N} n(d_i)[\log P(d_i) + \sum_{j=1}^{M} \frac{n(d_i, w_j)}{n(d_i)} \log \sum_{k=1}^{K} P(w_j|z_k)P(z_k|d_i)] \quad (3.12)$$

with respect to all probability mass functions. Here, $n(d_i) = \sum_j n(d_i, w_j)$ refers to the document length. Since the cardinality of the latent variable space is typically smaller than the number of the documents or the number of the terms in a collection, i.e., $K \ll \min(N, M)$, it acts as a bottleneck variable in predicting words. It is worth noting that an equivalent parameterization of the joint probability in Equation 3.11 can be obtained by:

$$P(d_i, w_j) = \sum_{k=1}^{K} P(z_k)P(d_i|z_k)P(w_j|z_k) \qquad (3.13)$$

which is perfectly symmetric in both entities, documents and words.

3.3.3   Model Fitting with the EM Algorithm

The standard procedure for maximum likelihood estimation in the latent variable model is the Expectation-Maximization (EM) algorithm. EM alternates in two steps: (i) an expectation (E) step where posterior probabilities are computed for the latent variables, based on the current estimates of the parameters; and (ii) a maximization (M) step, where parameters are updated based on the so-called expected complete data log-likelihood which depends on the posterior probabilities computed in the E-step.

For the E-step one simply applies Bayes' formula, e.g., in the parameterization of Equation 3.11, to obtain

$$P(z_k|d_i, w_j) = \frac{P(w_j|z_k)P(z_k|d_i)}{\sum_{l=1}^{K} P(w_j|z_l)P(z_l|d_i)} \qquad (3.14)$$

In the M-step one has to maximize the expected complete data log-likelihood $\mathbf{E}[\mathcal{L}^c]$. Since the trivial estimate $P(d_i) \propto n(d_i)$ can be carried out independently, the relevant part is given by

$$\mathbf{E}[\mathcal{L}^c] = \sum_{i=1}^{N} \sum_{j=1}^{M} n(d_i, w_j) \sum_{k=1}^{K} P(z_k|d_i, w_j) \log \left[P(w_j|z_k)P(z_k|d_i) \right] \qquad (3.15)$$

In order to take care of the normalization constraints, Equation 3.15 has to be augmented by appropriate Lagrange multiples τ_k and ρ_i,

$$\mathcal{H} = \mathbf{E}[\mathcal{L}^c] + \sum_{k=1}^{K} \tau_k (1 - \sum_{j=1}^{M} P(w_j|z_k)) + \sum_{i=1}^{N} \rho_i (1 - \sum_{k=1}^{K} P(z_k|d_i)) \qquad (3.16)$$

Maximization of $\mathcal{H}$ with respect to the probability mass functions leads to the following set of stationary equations

$$\sum_{i=1}^{N} n(d_i, w_j) P(z_k|d_i, w_j) - \tau_k P(w_j|z_k) = 0, 1 \leq j \leq M, 1 \leq k \leq K. \qquad (3.17)$$

$$\sum_{j=1}^{M} n(d_i, w_j) P(z_k|d_i, w_j) - \rho_i P(z_k|d_i) = 0, 1 \leq i \leq N, 1 \leq k \leq K. \qquad (3.18)$$

After eliminating the Lagrange multipliers, one obtains the M-step re-estimation equations

$$P(w_j|z_k) = \frac{\sum_{i=1}^{N} n(d_i, w_j) P(z_k|d_i, w_j)}{\sum_{m=1}^{M} \sum_{i=1}^{N} n(d_i, w_m) P(z_k|d_i, w_m)} \qquad (3.19)$$

$$P(z_k|d_i) = \frac{\sum_{j=1}^{M} n(d_i, w_j) P(z_k|d_i, w_j)}{n(d_i)} \qquad (3.20)$$

The E-step and M-step equations are alternated until a termination condition is met. This can be a convergence condition, but one may also use a technique known as *early stopping*. In early stopping one does not necessarily optimize until convergence, but instead stops updating the parameters once the performance on hold-out data is not improved. This is a standard procedure that can be used to avoid overfitting in the context of many iterative fitting methods, with EM being a special case.

3.3.4 Latent Probability Space and Probabilistic Latent Semantic Analysis

Consider the class-conditional probability mass functions $P(\bullet|z_k)$ over the vocabulary W which can be represented as points on the $M - 1$ dimensional simplex of all probability mass functions over W. Via its convex hull, this set of K points defines a $k - 1$ dimensional convex region $\mathcal{R} \equiv conv(P(\bullet|z_1), ..., P(\bullet|z_k))$ on the simplex (provided that they are in general positions). The modeling assumption expressed by Equation 3.11 is that all conditional probabilities $P(\bullet|d_i)$ for $1 \leq i \leq N$ are approximated by a convex combination of the K probability mass functions $P(\bullet|z_k)$. The mixing weights $P(z_k|d_i)$ are coordinates that uniquely define for each document a point within the convex region $\mathcal{R}$. This demonstrates that despite the discreteness of the introduced latent variables, a *continuous latent space* is obtained within the space of all probability mass functions over W. Since the dimensionality of the convex region $\mathcal{R}$ is $K - 1$ as opposed to $M - 1$ for the probability simplex, this can also be considered as the dimensionality reduction for the terms and $\mathcal{R}$ can be identified as a *probabilistic latent semantic space*. Each "direction" in the space corresponds to a particular context as quantified by $P(\bullet|z_k)$ and each document d_i participates in each context with a specific fraction $P(z_k|d_i)$. Note that since the aspect model is symmetric with respect to terms and documents, by reversing their roles one obtains a corresponding region $\mathcal{R}'$ in the simplex of all probability mass functions over D. Here each term w_j participates in each context with a fraction $P(z_k|w_j)$, i.e., the probability of an occurrence of w_j as part of the context z_k.

To stress this point and to clarify the relation to LSA, the aspect model as parameterized in Equation 3.13 is rewritten in matrix notion. Hence, define matrices by $\hat{U} = (P(d_i|z_k))_{i,k}$, $\hat{V} = (P(w_j|z_k))_{j,k}$, and $\hat{S} = diag(P(z_k))_k$. The joint probability model P can then be written as a matrix product $P = \hat{U}\hat{S}\hat{V}^T$. Comparing this decomposition with the SVD decomposition in LSA, one immediately points out the following interpretation of the concepts in linear algebra: (i) the weighted sum over outer products between rows of $\hat{U}$ and $\hat{V}$ reflects the conditional independence in pLSA; (ii) the K factors are seen to correspond to the mixture components of the aspect model; and (iii) the mixing proportions in pLSA substitute the singular values of the SVD in LSA. The crucial difference between pLSA and LSA, however, is the objective function utilized to determine the optimal decomposition/approximation. In

LSA, this is the L_2- or Frobenius norm, which corresponds to an implicit additive Gaussian noise assumption on (possibly transformed) counts. In contrast, pLSA relies on the likelihood function of the multinomial sampling and aims at an explicit maximization of the cross entropy of the Kullback-Leibler divergence between the empirical distribution and the model, which is different from any type of the squared deviation. On the modeling side this offers important advantages; for example, the mixture approximation P of the co-occurrence table is a well-defined probability distribution, and the factors have a clear probabilistic meaning in terms of the mixture component distributions. On the other hand, LSA does not define a properly normalized probability distribution, and even worse, $\tilde{N}$ may contain negative entries. In addition, the probabilistic approach can take advantage of the well-established statistical theory for model selection and complexity control, e.g., to determine the optimal number of latent space dimensions.

3.3.5   Model Overfitting and Tempered EM

The original model fitting technique using the EM algorithm has an over-fitting problem; in other words, its generalization capability is weak. Even if the performance on the training data is satisfactory, the performance on the testing data may still suffer substantially. One metric to assess the generalization performance of a model is called *perplexity*, which is a measure commonly used in language modeling. The perplexity is defined to be the log-averaged inverse probability on the unseen data, i.e.,

$$\mathcal{P} = \exp[-\frac{\sum_{i,j} n'(d_i, w_j) \log P(w_j|d_i)}{\sum_{i,j} n'(d_i, w_j)}] \tag{3.21}$$

where $n'(d_i, w_j)$ denotes the counts on hold-out or test data.

To derive conditions under which a generalization on the unseen data can be guaranteed is actually the fundamental problem of a statistical learning theory. One generalization of maximum likelihood for mixture models is known as *annealing* and is based on an entropic regularization term. The method is called *tempered expectation-maximization* (TEM) and is closely related to the *deterministic annealing* technique. The combination of deterministic annealing with the EM algorithm is the foundation basis of TEM.

The starting point of TEM is a derivation of the E-step based on an optimization principle. The EM procedure in latent variable models can be obtained by minimizing a common objective function — the (Helmholtz) *free energy* — which for the aspect model is given by

$$\mathcal{F}_\beta = -\beta \sum_{i=1}^{N} \sum_{j=1}^{M} n(d_i, w_j) \sum_{k=1}^{K} \tilde{P}(z_k; d_i, w_j) \log[P(d_i|z_k)P(w_j|z_k)P(z_k)]$$

$$= +\sum_{i=1}^{N} n(d_i) \sum_{k=1}^{K} \tilde{P}(z_k; d_i, w_j) \log \tilde{P}(z_k; d_i, w_j) \tag{3.22}$$

Here $\tilde{P}(z_k; d_i, w_j)$ are variational parameters which define a conditional distribution over $z_1, ..., z_K$ and β is a parameter which — in analogy to physical systems — is called the *inverse computational temperature*. Notice that the first contribution in Equation 3.22 is the negative expected log-likelihood scaled by β. Thus, in the case of $\tilde{P}(z_k; d_i, w_j) = P(z_k | d_i, w_j)$ minimizing $\mathcal{F}$ w.r.t. the parameters defining $P(d_i, w_j | z_k)$ amounts to the standard M-step in EM. In fact, it is straightforward to verify that the posteriors are obtained by minimizing $\mathcal{F}$ w.r.t. $\tilde{P}$ at $\beta = 1$. In general $\tilde{P}$ is determined by

$$\tilde{P}(z_k; d_i, w_j) = \frac{[P(z_k)P(d_i|z_k)P(w_j|z_k)]^\beta}{\sum_l [P(z_l)P(d_i|z_l)P(w_j|z_l)]^\beta} = \frac{[P(z_k|d_i)P(w_j|z_k)]^\beta}{\sum_l [P(z_l|d_i)P(w_j|z_l)]^\beta} \tag{3.23}$$

This shows that the effect of the entropy at $\beta < 1$ is to dampen the posterior probabilities such that they will be closer to the uniform distribution with decreasing β.

Somewhat contrary to the spirit of annealing as a continuation method, an "inverse" annealing strategy which first performs EM iterations and then decreases β until performance on the hold-out data deteriorates can be used. Compared with annealing, this may accelerate the model fitting procedure significantly. The TEM algorithm can be implemented in the following way:

1. Set $\beta \longleftarrow 1$ and perform EM with early stopping.

2. Decrease $\beta \longleftarrow \eta\beta$ (with $\eta < 1$) and perform one TEM iteration.

3. As long as the performance on hold-out data improves (non-negligibly), continue TEM iteration at this value of β; otherwise, goto step 2.

4. Perform stopping on β, i.e., stop when decreasing β does not yield further improvements.

3.4 Latent Dirichlet Allocation for Discrete Data Analysis

The Latent Dirichlet Allocation (LDA) is a statistical model for analyzing discrete data, initially proposed for document analysis. It offers a framework for understanding why certain words tend to occur together. Namely, it posits (in a simplification) that each document is a mixture of a small number of topics and that each word's creation is attributable to one of the document's topics. It is a graphical model for topic discovery developed by Blei, Ng, and Jordan [23] in 2003.

LDA is a generative language model which attempts to learn a set of topics and sets of words associated with each topic, so that each document may be viewed as a mixture of various topics. This is similar to pLSA, except that in LDA the topic distribution is assumed to have a Dirichlet prior. In practice, this results in more reasonable mixtures of topics in a document. It has been noted, however, that the pLSA model is equivalent to the LDA model under a uniform Dirichlet prior distribution [89].

For example, an LDA model might have topics "cat" and "dog". The "cat" topic has probabilities of generating various words: the words *tabby*, *kitten*, and, of course, *cat* will have high probabilities given this topic. The "dog" topic likewise has probabilities of generating words in which *puppy* and *dachshund* might have high probabilities. Words without special relevance, like *the*, will have roughly an even probability between classes (or can be placed into a separate category or even filtered out).

A document is generated by picking a distribution over topics (e.g., mostly about "dog", mostly about "cat", or a bit of both), and given this distribution, picking the topic of each specific word. Then words are generated given their topics. Notice that words are considered to be independent given the topics. This is the standard "bag of words" assumption, and makes the individual words exchangeable.

Learning the various distributions (the set of topics, their associated words' probabilities, the topic of each word, and the particular topic mixture of each document) is a problem of Bayesian inference, which can be carried out using the variational methods (or also with Markov Chain Monte Carlo methods, which tend to be quite slow in practice) [23]. LDA is typically used in language modeling for information retrieval.

3.4.1 Latent Dirichlet Allocation

While the pLSA described in the last section is very useful toward probabilistic modeling of multimedia data units, it is argued to be incomplete in that it provides no probabilistic model at the level of the documents. In pLSA, each document is represented as a list of numbers (the mixing proportions for topics), and there is no generative probabilistic model for these numbers. This leads to two major problems: (1) the number of parameters in the model grows linearly with the size of the corpus, which leads to serious problems with overfitting; and (2) it is not clear how to assign a probability to a document outside a training set.

LDA is a truly generative probabilistic model that not only assigns probabilities to documents of a training set, but also assigns probabilities to other documents not in the training set. The basic idea is that documents are represented as random mixtures over latent topics, where each topic is characterized by a distribution over words. LDA assumes the following generative process for each document w in a corpus D:

1. Choose $N \sim Poisson(\xi)$.

2. Choose $\theta \sim Dir(\alpha)$.

3. For each of the N words w_n:

   Choose a topic $z_n \sim Multinomial(\theta)$.

   Choose a word w_n from $p(w_n | z_n, \boldsymbol{\beta})$, a multinomial probability conditioned on the topic z_n.

where $Poisson(\xi)$, $Dir(\alpha)$, and $Multinomial(\theta)$ denote a Poisson, a Dirichlet, and a multinomial distribution with parameters ξ, α, and θ, respectively. Several simplifying assumptions are made in this basic model. First, the dimensionality k of the Dirichlet distribution (and thus the dimensionality of the topic variable z) is assumed known and fixed. Second, the word probabilities are parameterized by a $k \times V$ matrix $\boldsymbol{\beta}$ where $\beta_{ij} = p(w^j = 1 | z^i = 1)$, which is treated as a fixed quantity that is to be estimated. Finally, the Poisson assumption is not critical to the modeling, and a more realistic document length distribution can be used as needed. Furthermore, note that N is independent of all the other data generation variables (θ and z). It is thus an ancillary variable.

A k-dimensional Dirichlet random variable θ can take values in the $(k-1)$-simplex (a k-dimensional vector θ lies in the $(k-1)$-simplex if $\theta_j \geq 0, \sum_{j=1}^{k} \theta_j = 1$), and has the following density on this simplex:

$$p(\theta|\alpha) = \frac{\Gamma(\sum_{i=1}^{k} \alpha_i)}{\prod_{i=1}^{k} \Gamma(\alpha_i)} \theta_1^{\alpha_1 - 1} \ldots \theta_k^{\alpha_k - 1} \qquad (3.24)$$

where the parameter α is a k-dimensional vector with components $\alpha_i > 0$, and where $\Gamma(x)$ is the Gamma function. The Dirichlet is a convenient distribution on the simplex — it is in the exponential family, has finite dimensional sufficient statistics, and is conjugate to the multinomial distribution. These properties facilitate the development of inference and parameter estimation algorithms for LDA.

Given the parameters α and β, the joint distribution of a topic mixture θ, a set of N topics $\mathbf{z}$, and a set of N words $\mathbf{w}$ is given by:

$$p(\theta, \mathbf{z}, \mathbf{w} | \alpha, \beta) = p(\theta|\alpha) \prod_{n=1}^{N} p(z_n|\theta) p(w_n|z_n, \beta) \qquad (3.25)$$

where $p(z_n|\theta)$ is simply θ_i for the unique i such that $z_n^i = 1$. Integrating over θ and summing over $\mathbf{z}$, we obtain the marginal distribution of a document:

$$p(\mathbf{w}|\alpha, \beta) = \int p(\theta|\alpha) \left(\prod_{n=1}^{N} \sum_{z_n} p(z_n|\theta) p(w_n|z_n, \beta) \right) d\theta \qquad (3.26)$$

Finally, taking the product of the marginal probabilities of single documents d, we obtain the probability of a corpus D with M documents:

$$p(D|\alpha, \beta) = \prod_{d=1}^{M} \int p(\theta_d|\alpha) \left(\prod_{n=1}^{N_d} \sum_{z_{dn}} p(z_{dn}|\theta_d) p(w_{dn}|z_{dn}, \beta) \right) d\theta_d \qquad (3.27)$$

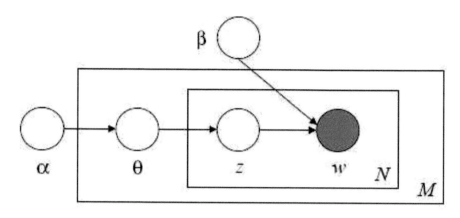

FIGURE 3.2: Graphical model representation of LDA. The boxes are "plates" representing replicates. The outer plate represents documents, while the inner plate represents the repeated choice of topics and words within a document.

The LDA model is represented as a probabilistic graphical model in Figure 3.2. As the figure indicates clearly, there are three levels to the LDA representation. The parameters α and β are corpus-level parameters, assumed to be sampled once in the process of generating a corpus. The variables θ_d are document-level variables, sampled once per document. Finally, the variables z_{dn} and w_{dn} are word-level variables and are sampled once for each word in each document.

It is important to distinguish LDA from a simple Dirichlet-multinomial clustering model. A classical clustering model would involve a two-level model in which a Dirichlet is sampled once for a corpus, a multinomial clustering variable is selected once for each document in the corpus, and a set of words is selected for the document conditional on the cluster variable. As with many clustering models, such a model restricts a document to being associated with a single topic. LDA, on the other hand, involves three levels, and notably the topic node is sampled repeatedly within the document. Under this model, documents can be associated with multiple topics.

3.4.2 Relationship to Other Latent Variable Models

In this section we compare LDA with simpler latent variable models — the unigram model, a mixture of unigrams, and the pLSA model. Furthermore, we present a unified geometric interpretation of these models which highlights their key differences and similarities.

1. Unigram model

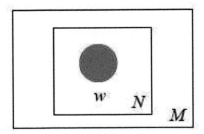

FIGURE 3.3: Graphical model representation of unigram model of discrete data.

Under the unigram model, the words of every document are drawn independently from a single multinomial distribution:

$$p(\mathbf{w}) = \prod_{n=1}^{N} p(w_n)$$

This is illustrated in the graphical model in Figure 3.3.

2. Mixture of unigrams

   If we augment the unigram model with a discrete random topic variable z (Figure 3.4), we obtain a mixture of unigrams model. Under this mixture model, each document is generated by first choosing a topic z and then generating N words independently from the conditional multinomial $p(w|z)$. The probability of a document is:

$$p(\mathbf{w}) = \sum_{z} p(z) \prod_{n=1}^{N} p(w_n|z)$$

   When estimated from a corpus, the word distributions can be viewed as representations of topics under the assumption that each document exhibits exactly one topic. This assumption is often too limiting to effectively model a large collection of documents. In contrast, the LDA model allows documents to exhibit multiple topics to different degrees. This is achieved at a cost of just one additional parameter: there are $k-1$ parameters associated with $p(z)$ in the mixture of unigrams, versus the k parameters associated with $p(\theta|\alpha)$ in LDA.

3. Probabilistic latent semantic analysis

   Probabilistic latent semantic analysis (pLSA), introduced in Section 3.3 is another widely used document model. The pLSA model, illustrated

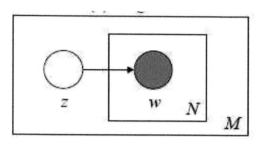

FIGURE 3.4: Graphical model representation of mixture of unigrams model of discrete data.

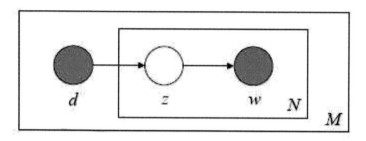

FIGURE 3.5: Graphical model representation of pLSI/aspect model of discrete data.

in Figure 3.5, posits that a document label d and a word w_n are conditionally independent given an unobserved topic z:

$$p(d, w_n) = p(d) \sum_z p(w_n|z)p(z|d)$$

.

The pLSA model attempts to relax the simplifying assumption made in the mixture of unigrams model that each document is generated from only one topic. In a sense, it does capture the possibility that a document may contain multiple topics since $p(z|d)$ serves as the mixture weights of the topics for a particular document d. However, it is important to note that d is a dummy index into the list of documents in the training set. Thus, d is a multinomial random variable with as many possible values as there are in the training documents, and the model learns the topic mixtures $p(z|d)$ only for those documents on which it is trained. For this reason, pLSA is not a well-defined generative model of documents; there is no natural way to use it to assign a probability to a previously unseen document. A further difficulty with pLSA, which also stems from the use of a distribution indexed by the training documents, is that the number of the parameters which must be estimated grows linearly with the number of the training documents. The parameters for a k-topic pLSA model are k multinomial distributions in size V and M mixtures over the k hidden topics. This gives $kV + kM$ parameters and therefore a linear growth in M. The linear growth in parameters suggests that the model is prone to overfitting, and, empirically, overfitting is indeed a serious problem. In practice, a tempering heuristic is used to smooth the parameters of the model for an acceptable predictive performance. It has been shown, however, that overfitting can occur even when tempering is used. LDA overcomes both of these problems by treating the topic mixture weights as a k-parameter hidden random variable rather than a large set of individual parameters which are explicitly linked to the training set. As described above, LDA is a well-defined generative model and generalizes easily to new documents. Furthermore, the $k + kV$ parameters in a k-topic LDA model do not grow with the size of the training corpus. In consequence, LDA does not suffer from the same overfitting issues as pLSA.

3.4.3   Inference in LDA

We have described the motivation behind LDA and have illustrated its conceptual advantages over other latent topic models. In this section, we turn our attention to procedures for inference and parameter estimation under LDA.

The key inferential problem that we need to solve in order to use LDA is that of computing the posterior distribution of the hidden variables given a document:

$$p(\theta, \mathbf{z} | \mathbf{w}, \alpha, \beta) = \frac{p(\theta, \mathbf{z}, \mathbf{w} | \alpha, \beta)}{p(\mathbf{w} | \alpha, \beta)}$$

Unfortunately, this distribution is intractable to compute in general. Indeed, to normalize the distribution we marginalize over the hidden variables and write Equation 3.26 in terms of the model parameters:

$$p(\mathbf{w} | \alpha, \beta) = \frac{\Gamma(\sum_i \alpha_i)}{\prod_i \Gamma(\alpha_i)} \int (\prod_{i=1}^{k} \theta_i^{\alpha_i - 1}) (\prod_{n=1}^{N} \sum_{i=1}^{k} \prod_{j=1}^{V} (\theta_i \beta_{ij})^{w_n^j}) d\theta$$

a function which is intractable due to the coupling between θ and β in the summation over latent topics. It has been shown that this function is an expectation under a particular extension to the Dirichlet distribution which can be represented with special hypergeometric functions. It has been used in a Bayesian context for censored discrete data to represent the posterior on θ which, in that setting, is a random parameter.

Although the posterior distribution is intractable for exact inference, a wide variety of approximate inference algorithms can be considered for LDA, including Laplace approximation, variational approximation, and Markov chain

Monte Carlo. In this section we describe a simple convexity-based variational algorithm for inference in LDA.

The basic idea of convexity-based variational inference is to obtain an adjustable lower bound on the log likelihood. Essentially, one considers a family of lower bounds, indexed by a set of *variational parameters*. The variational parameters are chosen by an optimization procedure that attempts to find the tightest possible lower bound.

A simple way to obtain a tractable family of lower bounds is to consider simple modifications of the original graphical model in which some of the edges and nodes are removed. The problematic coupling between θ and β arises due to the edges between θ, $\mathbf{z}$, and $\mathbf{w}$. By dropping these edges and the $\mathbf{w}$ nodes, and endowing the resulting simplified graphical model with free variational parameters, we obtain a family of distributions on the latent variables. This family is characterized by the following variational distribution:

$$p(\theta, \mathbf{z}|\gamma, \phi) = p(\theta|\gamma) \prod_{n=1}^{N} p(z_n|\phi_n) \qquad (3.28)$$

where the Dirichlet parameter γ and the multinomial parameters $(\phi_1, ..., \phi_N)$ are the free variational parameters.

We summarize the variational inference procedure in Algorithm 1, with appropriate starting points for γ and ϕ_n. From the pseudocode it is clear that each iteration of the variational inference for LDA requires $O((N+1)k)$ operations. Empirically, we find that the number of iterations required for a single document is in the order of the number of words in the document. This yields a total number of operations roughly in the order of $N^2 k$.

3.4.4   Parameter Estimation in LDA

In this section we present an empirical Bayes method for parameter estimation in the LDA model. In particular, given a corpus of documents $D = \{\mathbf{w_1}, \mathbf{w_2}, ...\mathbf{w_M}\}$, we wish to find parameters α and β that maximize the (marginal) log likelihood of the data:

$$L(\alpha, \beta) = \sum_{d=1}^{M} \log p(\mathbf{w_d}|\alpha, \beta)$$

As described above, the quantity $p(\mathbf{w}|\alpha, \beta)$ cannot be computed tractably. However, the variational inference provides us with a tractable lower bound on the log likelihood, a bound which we can maximize with respect to α and β. We can thus find approximate empirical Bayes estimates for the LDA model via an alternating *variational EM* procedure that maximizes a lower bound with respect to the variational parameters γ and ϕ, and then, for the fixed values of the variational parameters, maximizes the lower bound with respect to the model parameters α and β.

Algorithm 1 A variational inference algorithm for LDA

Input: A corpus of documents with N words w_n and k topics (i.e., clusters)
Output: Parameters ϕ and γ
Method:

1: Initialize $t = 0$
2: Initialize ${\phi_{ni}}^t = \frac{1}{k}$ for all i and n
3: Initialize ${\gamma_i}^t = \alpha_i + \frac{N}{k}$ for all i
4: **repeat**
5:     **for** $n = 1$ to N **do**
6:       **for** $i = 1$ to k **do**
7:         ${\phi_{ni}}^{t+1} = \beta_{iw_n} \exp(\Phi({\gamma_i}^t))$
8:       **end for**
9:       Normalize ${\phi_n}^{t+1}$ to sum to 1
10:     **end for**
11:     $\gamma^{t+1} = \alpha + \sum_{n=1}^{N} {\phi_n}^{t+1}$
12:     $t = t + 1$
13: **until** Convergence

There is a detailed derivation of the variational EM algorithm for LDA [23]. The derivation yields the following iterative algorithm:

1. (E-step) For each document, find the optimizing values of the variational parameters $\{\gamma_d^*, \phi_d^* : d \in D\}$. This is done as described in the previous section.

2. (M-step) Maximize the resulting lower bound on the log likelihood with respect to the model parameters α and β. This corresponds to finding the maximum likelihood estimates with the expected sufficient statistics for each document under the approximate posterior which is computed in the E-step.

These two steps are repeated until the lower bound on the log likelihood converges. In addition, the M-step update for the conditional multinomial parameter β can be written out analytically:

$$\beta_{ij} = \sum_{d=1}^{M} \sum_{n=1}^{N_d} \phi_{dni}^* w_{dn}^j$$

It is also shown that the M-step update for Dirichlet parameter α can be implemented using an efficient Newton-Raphson method in which the Hessian is inverted in a linear time.

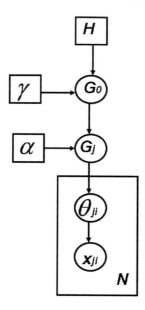

FIGURE 3.6: Graphical model representation of the Hierarchical Dirichlet Process of discrete data.

3.5   Hierarchical Dirichlet Process

indexhierarchical Dirichlet process

All the proposed language models introduced so far have a fundamental assumption that the number of the topics in the data corpus must be given in advance. Given the fact that all the Bayesian models can be developed into a hierarchy, recently Teh et al have proposed a nonparametric hierarchical Bayesian model called the *Hierarchical Dirichlet Process*, abbreviated as HDP [203]. The advantage of HDP in comparison with the existing latent models is that HDP is capable of automatically determining the number of topics or clusters and sharing the mixture components across topics.

Specifically, HDP is based on Dirichlet process mixture models where it is assumed that the data corpora have different groups and each group is associated with a mixture model, with all the groups sharing the same set of mixture components. With this assumption, the number of clusters can be left as open-ended. Consequently, HDP is ideal for multi-tasking learning or learning to learn. When there is only one group, HDP is reduced to LDA. Figure 3.6 shows the graphical model for HDP. The corresponding generative process is given as follows.

1. A random measure G_0 is drawn from a Dirichlet process $\mathscr{DP}$ [24, 258,

158] parameterized by concentration parameter α and base probability measure H:

$$G_0 | \gamma, H \sim \mathscr{DP}(\gamma, H). \tag{3.29}$$

G_0 can be constructed by the stick-breaking process [77, 19], i.e.,

$$G_0 = \sum_{k=1}^{\infty} \pi_k \delta_{\phi_k}$$

$$\begin{aligned} \pi'_k | \gamma, H &\sim Beta(1, \gamma) \\ \phi_k | \gamma, H &\sim H \\ \pi_k &= \pi'_k \prod_{l=1}^{k-1}(1 - \pi'_l) \\ G_0 &= \sum_{k=1}^{\infty} \pi_k \delta_{\phi_k} \end{aligned} \tag{3.30}$$

where δ_ϕ denotes a probability measure concentrated on ϕ, π is the weight set $\{\pi_k\}_{k=1}^{\infty}$, and $Beta$ is a Beta distribution.

2. A random probability measure G_j^d for each document j is drawn from a Dirichlet process with concentration parameter α and base probability measure G_0:

$$G_j^d | \alpha, G_0 \sim \mathscr{DP}(\alpha, G_0). \tag{3.31}$$

In this case, G_j^d, which is a prior distribution of all the words in document j, shares the mixture components $\{\phi_k\}_{k=1}^{\infty}$ with G_0. Namely, G_j^d can be written as $G_j^d = \sum_{k=1}^{\infty} \pi_{jk} \delta_{\phi_k}$.

3. A topic θ_{ji} for each word i in document j is drawn from G_j^d such that θ_{ji} is sampled as one of $\{\phi_k\}_{k=1}^{\infty}$.

4. Each word i in document j, i.e., x_{ji}, is drawn from a likelihood distribution $F(x_{ji}|\theta_{ji})$.

3.6 Applications in Multimedia Data Mining

Ever since the idea of the latent topic (or the latent concept) discovery from a document corpus reflected by LDA [23] or HDP [203] or the related pLSA [101] was published, these language models have succeeded substantially in text or information retrieval. Due to this success, a number of applications of these language models to multimedia data mining have been reported in the literature. Noticeable examples include using LDA to discover objects in image collections [191, 181, 33, 213], using pLSI to discover semantic concepts in image collections [240, 243], using LDA to classify scene image categories [73, 74, 76], using pLSI to learn image annotation [155, 245, 246], and using LDA to understand the activities and interactions in surveillance video [214].

Since models such as LDA, pLSI, and HDP are originally proposed for text retrieval, the applications of these language models to multimedia data immediately lead to two different but related issues. The first issue is that in text data, each word is naturally and distinctly presented in the language vocabulary, and each document is also clearly represented as a collection of words; however, there is no clear correspondence in multimedia data for the concepts of words and documents. Consequently, how to appropriately represent a multimedia word and/or a multimedia document in multimedia data has become a non-trivial issue. In the current literature of multimedia data mining, a multimedia word is typically represented either as a segmented unit in the original multimedia data space (e.g., an image patch after a segmentation of an image [213]) or as a segmented unit in a transformed feature space (e.g., as a unit of a quantized motion feature space [214]); similarly, a multimedia document may be represented as a certain partition of the multimedia data, such as a part of an image [213], or an image [191], or a video clip [214].

The second and more important issue is that the original language models are based on the fundamental assumption that a document is simply a bag of words. However, in multimedia data, often there is a strong spatial correlation between the multimedia words in a multimedia document, such as the neighboring pixels or regions in an image, or related video frames of a video stream. In order to make those language models work effectively, we must incorporate the spatial information into these models. Consequently, in the recent literature, variations of these language models are developed specifically tailored to the specific multimedia data mining applications. For example, Cao and Fei-Fei propose the Spatially Coherent Latent Topic Model (Spatial-LTM) [33] and Wang and Grimson propose the Spatial Latent Dirichlet Allocation (SLDA) [213]. To further model the temporal correlation for temporal data or time-series data, Teh et al [203] further propose a hierarchical Bayesian model that is a combination of the HDP model and the Hidden Markov Model (HMM) called HDP-HMM for automatic topic discovery and clustering, and have proven that the infinite Hidden Markov Model (iHMM) [17], based on the coupled urn model, is equivalent to an HDP-HMM.

3.7 Support Vector Machines

The *support vector machine* (SVM) is a supervised learning method typically used for classification and regression. SVMs are generalized linear classifiers. SVMs attempt to minimize the classification error through maximizing the geometric margin between classes. In this sense, SVMs are also called the *maximum margin classifiers*.

As a typical representation in classification, data points are represented as feature vectors in a feature space. A support vector machine maps these input vectors to a higher-dimensional space such that a separating hyperplane may be constructed between classes; this separating hyperplane may be constructed in such a way that the two parallel boundary hyperplanes to the separating hyperplane are constructed on each side of the hyperplane that the distance between these two boundary hyperplanes is maximized, where a boundary hyperplane is the hyperplane that passes at least one data point of the class and all the other data points of the class are located in the other side of the boundary hyperplane across the parallel separating hyperplane. Thus, the separating hyperplane is the hyperplane that maximizes the distance between the two parallel boundary hyperplanes. Presumably, the larger the margin or distance between these parallel boundary hyperplanes, the better the generalization error of the classifier is.

Let us first focus on the simplest scenario of classification — the two-class classification. Each data point is represented as a p-dimensional vector in a p-dimensional Euclidean feature space. Each of these data points is in only one of the two classes. We are interested in whether we can separate these data points of the two classes with a $p-1$ dimensional hyperplane. This is a standard problem of linear classifiers. There are many linear classifiers as solutions to this problem. However, we are particularly interested in determining whether we can achieve maximum separation (i.e., the maximum margin) between the two classes. By the maximum margin, we mean that we determine the separating hyperplane between the two classes such that the distance from the separating hyperplane to the *nearest* data point in either of the classes is maximized. That is equivalent to say that the distance between the two parallel boundary hyperplanes to the separating hyperplane is maximized. If such a separating hyperplane exists, it is clearly of interest and is called the *maximum-margin hyperplane*; correspondingly, such a linear classifier is called a *maximum margin classifier*. Figure 3.7 illustrates the different separating hyperplanes on a two-class data set.

We consider data points of the form $\{(\mathbf{x_1}, c_1), (\mathbf{x_2}, c_2), ..., (\mathbf{x_n}, c_n)\}$, where the c_i is either 1 or -1, a constant denoting the class to which the point $\mathbf{x_i}$ belongs. Each is a p-dimensional real vector and may be normalized into the range of [0,1] or [-1,1]. The scaling is important to guard against variables with larger variances that might otherwise dominate the classification. At present we take this data set as the training data; the training data set represents the correct classification that we would like an SVM to eventually perform, by means of separating the data with a hyperplane, in the form

$$\mathbf{w} \cdot \mathbf{x} - b = 0$$

The vector $\mathbf{w}$ is perpendicular to the separating hyperplane. With the offset parameter b, we are allowed to increase the margin, as otherwise the hyperplane must pass through the origin, restricting the solution. Since we are interested in the maximum margin, we are interested in those data points

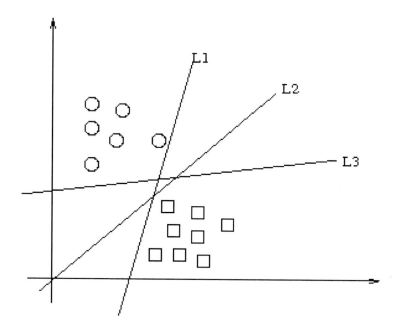

FIGURE 3.7: Different separating hyperplanes on a two-class data set.

closer or touch the parallel boundary hyperplanes to the separating hyperplane between the two classes. It is easy to show that these parallel boundary hyperplanes are described by equations (through scaling w and b) $\mathbf{w} \cdot \mathbf{x} - b = 1$ and $\mathbf{w} \cdot \mathbf{x} - b = -1$. If the training data are linearly separable, we select these hyperplanes such that there are no points between them and then try to maximize their distance. By the geometry, we find the distance between the hyperplanes is $2/|\mathbf{w}|$ (as shown in Figure 3.8); consequently, we attempt to minimize $|\mathbf{w}|$. To exclude any data points between the two parallel boundary hyperplanes, we must ensure that for all i, either $\mathbf{w} \cdot \mathbf{x} - b \geq 1$ or $\mathbf{w} \cdot \mathbf{x} - b \leq -1$. This can be rewritten as:

$$c_i(\mathbf{w} \cdot \mathbf{x_i} - b) \geq 1, 1 \leq i \leq n \qquad (3.32)$$

where those data points $\mathbf{x}$ that make the inequality Equation 3.32 an equality are called *support vectors*. Geometrically, support vectors are those data points that are located on either of the two parallel boundary hyperplanes.

The problem now is to minimize $|\mathbf{w}|$ subject to the constraint (3.32). This is a quadratic programming (QP) optimization problem. Further, the problem is to minimize $(1/2)||\mathbf{w}||^2$, subject to Equation 3.32. Writing this classification problem in its dual form reveals that the classification solution is only determined by the support vectors, i.e., the training data that lie on the margin. The dual of the SVM is:

$$\max \sum_{i=1}^{n} \alpha_i - \sum_{i,j} \alpha_i \alpha_j c_i c_j \mathbf{x_i}^T \mathbf{x_j} \qquad (3.33)$$

subject to $\alpha_i \geq 0$, where the α terms constitute a dual representation for the weight vector in terms of the training set:

$$\mathbf{w} = \sum_{i}^{n} \alpha_i c_i \mathbf{x_i} \qquad (3.34)$$

At this time it is assumed that there always exists a separating hyperplane that perfectly separates the given training samples of the two classes. What if there are errors in the training data such that there is no such perfect separating hyperplane existing? Cortes and Vapnik have proposed a modified maximum margin method that allows for mislabeled examples [53]. If there exists no hyperplane that can split the +1 and −1 examples, this method selects a hyperplane that splits the examples as cleanly as possible, while still maximizing the distance to the nearest cleanly split examples. This work further promotes the understanding of SVM. This proposed method is called *soft-margin SVM*. The soft-margin SVM introduces slack variables, ξ_i, which measure the degree of misclassification of the datum $\mathbf{x_i}$:

$$c_i(\mathbf{w} \cdot \mathbf{x_i} - b) \geq 1 - \xi_i, 1 \leq i \leq n \qquad (3.35)$$

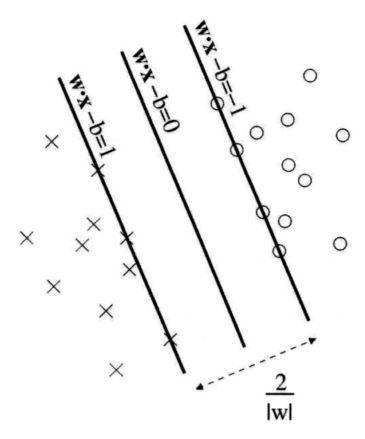

FIGURE 3.8: Maximum-margin hyperplanes for an SVM trained with samples of two classes. Samples on the boundary hyperplanes are called the support vectors.

The objective function is then increased by a function which penalizes non-zero ξ_i, and the optimization becomes a trade-off between a large margin and a small error penalty. If the penalty function is linear, the objective function now becomes

$$\min ||\mathbf{w}||^2 + C \sum_i^n \xi_i \qquad (3.36)$$

such that the constraint Equation 3.35 holds true. The minimization in Equation 3.36 may be solved using Lagrange multipliers. The obvious advantage of a linear penalty function is that the slack variables vanish from the dual problem, with the constant C appearing only as an additional constraint on the Lagrange multipliers. Non-linear penalty functions are also used in the literature, particularly to reduce the effect of outliers on the classifier; however, typically the problem becomes non-convex, and thus it is considerably more difficult to find a global solution.

The original optimal hyperplane algorithm developed by Vapnik and Lerner [208] was a linear classifier. Later Boser, Guyon, and Vapnik addressed the non-linear classifiers by applying the kernel trick (originally proposed by Aizerman et al [7]) to the maximum-margin hyperplanes [27]. The resulting algorithm was formally similar to the linear solution, except that every dot product was replaced with a non-linear kernel function. This allows the algorithm to fit the maximum-margin separating hyperplane in the transformed feature space. The transformation may be non-linear and the transformed space may be high-dimensional; consequently, the classifier becomes a separating hyperplane in a higher-dimensional feature space but at the same time it is non-linear in the original feature space, also.

If the kernel used is a Gaussian radial basis function, the corresponding feature space is a Hilbert space of infinite dimension. Maximum margin classifiers thus become well regularized. Consequently, the infinite dimension does not spoil the results. Commonly used kernels include:

1. Polynomial (homogeneous): $k(\mathbf{x}, \mathbf{x}') = (\mathbf{x} \cdot \mathbf{x}')^d$

2. Polynomial (inhomogeneous): $k(\mathbf{x}, \mathbf{x}') = (\mathbf{x} \cdot \mathbf{x}' + 1)^d$

3. Radial Basis Function: $k(\mathbf{x}, \mathbf{x}') = \exp(-\gamma ||\mathbf{x} - \mathbf{x}'||^2)$ for $\gamma > 0$

4. Gaussian Radial basis function: $k(\mathbf{x}, \mathbf{x}') = \exp(\frac{||\mathbf{x} - \mathbf{x}'||^2}{2\sigma^2})$

5. Sigmoid: $k(\mathbf{x}, \mathbf{x}') = \tanh(k\mathbf{x} \cdot \mathbf{x}' + c)$, for some (not every) $k > 0$ and $c < 0$

SVMs were also proposed for regression by Vapnik et al [65]. This method is called *support vector regression* (SVR). As we have shown above, the classic support vector classification only depends on a subset of the training data, i.e., the support vectors, as the cost function does not care at all about the training data that lie beyond the margin. Correspondingly, SVR only depends on a subset of the training data, because the cost function also ignores any training data that are close (within a threshold ε) to the model prediction.

The parameters of the maximum-margin hyperplane are obtained by solving the optimization problem. In the literature there are several specialized algorithms for quickly solving the QP problem that arises from the SVMs; most of the solutions use heuristics for breaking the problem down into smaller, more manageable subproblems. A commonly used method for solving the QP problem is Platt's SMO algorithm [169], which breaks the original problem down into 2-dimensional subproblems that may be solved analytically, eliminating the need for a numerical optimization algorithm such as the conjugate gradient methods. Recent work includes the fast training of SVM such as Joachims [120], which gives a cutting plane algorithm for training an SVM. This is the first algorithm that optimizes the traditional hinge-loss SVM formulation in linear time in the size of the training data (where the training data are represented in the sparse format with zero-valued attributes not included). The software is available in SVMPerf, which is a freely downloadable off-the-shelf package [3].

For the case of multiple-classes classification, there are four commonly used approaches to extending the two-classes (called binary-classes) SVM classification method introduced above to the multiple-classes SVMs. Assume that there are n classes in the original n-classes classification problem, and that each of the n classes has N data samples. Below we use $O(a, b)$ to denote the two-classes SVM training complexity with the numbers of the data samples in the two classes are a and b, respectively.

- *One-against-one*: This is the most straightforward solution. For any pair of classes among the n classes, we apply a two-classes SVM. Then we need to simultaneously solve for a set of $\frac{n(n-1)}{2}$ two-classes classification problems. The final result may be obtained through statistical voting. Thus, for this approach, the total training complexity is $\frac{n(n-1)}{2}O(N, N)$, and it requires $\frac{n(n-1)}{2}$ total times of testing.

- *One-against-all*: In this approach, for each of the n classes, it is taken to be classified against the rest $n-1$ classes. Then we need to solve for a set of n two-classes classification problems. The final result may be obtained through statistical voting. Thus, for this approach, the total training complexity is $nO(N, (n-1)N)$, and it requires $n-1$ total times of testing.

- *Top-down binary tree*: In this approach, initially all the n classes are considered as a single group. A recursive splitting is repeatedly applied to the group to split into two classes and the two-classes SVM is applied until the test pattern is assigned to a final class. Thus, this approach requires $\sum_{i=1}^{\log_2 n} 2^{i-1}O(\frac{nN}{2}, \frac{nN}{2})$ total training complexity, and $\log_2 n$ total times of testing.

- *Bottom-up binary tree*: In this approach, a pair-wise SVM classification is applied initially to the whole group of the n classes and then recursively among the survivors until the test pattern is assigned to the final class. Thus, this approach requires $\frac{n(n-1)}{2}O(N,N)$ total training complexity, and $n-1$ total times of testing.

More extensive discussions on the solutions to the multiple-classes SVMs may be found in the literature [148].

3.8 Maximum Margin Learning for Structured Output Space

SVMs as the maximum margin classifiers are originally proposed for solving the classic classification problem. For a classic classification problem, the output space has the following two properties. First, the number of classes is finite where this number is typically a very limited integer. Second and more importantly, the classes are exclusive, and for each input data object, there is one but only one output class that maps to the input data object. However, in many multimedia data mining applications, the output space no longer satisfies these two properties. In other words, in the output space, classes are not exclusive with dependency among the classes such that an input data object may belong to multiple classes, and the number of the output classes may no longer be a very limited integer anymore. We call this type of output space a *structured space*. Examples of the classification problems with structured output space include machine translation in which both the input space and the output space are the specific language vocabulary spaces which are the structured space where words have the dependency in the translated meanings such that one input word may map to different words in the output space; learning the parsing tree where the input space is a language vocabulary space and the output space is a parsing tree which is another structured space; and the image annotation problem where the input space is a collection of images and the output space is a language vocabulary space which once again is a structured space.

To exploit the dependency relationships for a structured space, we need to learn the relationship among the dependency explicitly. Given two domains $\mathcal{X}$ and $\mathcal{Y}$ and an input $\mathbf{x} \in \mathcal{X}$ and an output $\mathbf{y} \in \mathcal{Y}$ as structures, the learning problem is therefore formulated as finding a function $f : \mathcal{X} \times \mathcal{Y} \to \mathbb{R}$ such that

$$\hat{\mathbf{y}} = \arg\max_{\mathbf{y} \in \mathcal{Y}} f(\mathbf{x}, \mathbf{y}) \tag{3.37}$$

is the desired output for the input $\mathbf{x}$.

To demonstrate how to solve for the maximum margin learning problem in a structured output space, we use the image annotation problem as an example [95]. Assume that the image database consists of a set of instances $S = \{(I_i, W_i)\}_{i=1}^{L}$, where each instance consists of an image object I_i and the corresponding annotation word set W_i. Note here that since the output space is a structured one, one input instance (here an image object) may be mapped to multiple words instead of just a single, unique class (i.e., a word). Here we assume that each image is partitioned into a set of blocks, and we use each block as an image object. Thus, an image can be represented by a set of such blocks. Further, each image block is represented as a feature vector in an image feature space. Consequently, an image is represented as a set of feature vectors in the feature space. A clustering algorithm is then applied to the whole feature space to group similar feature vectors together. The centroid of a cluster represents a visual representative (we refer to it as VRep here) in the image space. In Figure 3.9, there are two VReps, *water* and *duck in the water*. The corresponding annotation word set can be easily obtained for each VRep. Consequently, the image database becomes the VRep-word pairs $S = \{(\mathbf{x}_i, \mathbf{y}_i)\}_{i=1}^{n}$, where n is the number of the clusters, $\mathbf{x}_i$ is a VRep object, and $\mathbf{y}_i$ is the word annotation set corresponding to this VRep object. Another simple method to obtain the VRep-word pairs is that we randomly select some images from the image database and each image is viewed as a VRep.

Suppose that there are W distinct annotation words. An arbitrary subset of annotation words is represented by the binary vector $\bar{\mathbf{y}}$ whose length is W; the j-th component $\bar{\mathbf{y}}_j = 1$ if the j-th word occurs in this subset, and 0 otherwise. All possible binary vectors form the word space $\mathcal{Y}$. We use $\mathbf{w}_j$ to denote the j-th word in the whole word set. We use $\mathbf{x}$ to denote an arbitrary vector in the feature space. Figure 3.9 shows an illustrative example in which the original image is annotated by *duck* and *water* which are represented by a binary vector. There are two VReps after the clustering, and each has a different annotation. In the word space, a word may be related to other words. For example, *duck* and *water* are related to each other because *water* is more likely to occur when *duck* is one of the annotation words. Consequently, the annotation word space is a structured output space where the elements are interdependent.

The relationship between the input example VRep $\mathbf{x}$ and an arbitrary output $\bar{\mathbf{y}}$ is represented as the joint feature mapping $\Phi(\mathbf{x}, \bar{\mathbf{y}})$, $\Phi : \mathcal{X} \times \mathcal{Y} \rightarrow \mathbb{R}^d$ where d is the dimension of the joint feature space and $\mathbb{R}$ means a real space. It can be expressed as a linear combination of the joint feature mapping between $\mathbf{x}$ and all the unit vectors. That is,

$$\Phi(\mathbf{x}, \bar{\mathbf{y}}) = \sum_{j=1}^{W} \bar{\mathbf{y}}_j \Phi(\mathbf{x}, \mathbf{e}_j)$$

where $\mathbf{e}_j$ is the j-th unit vector. The score between $\mathbf{x}$ and $\bar{\mathbf{y}}$ can be expressed as a linear combination of each component in the joint feature representation: $f(\mathbf{x}, \bar{\mathbf{y}}) = \langle \boldsymbol{\alpha}, \Phi(\mathbf{x}, \bar{\mathbf{y}}) \rangle$. Then the learning task is to find the optimal weight

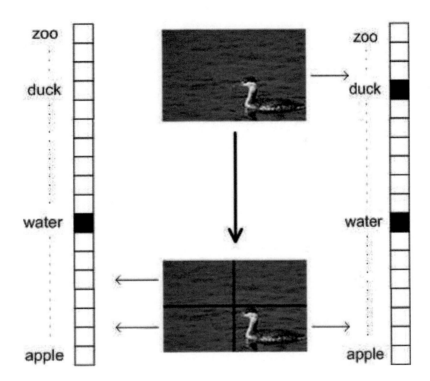

FIGURE 3.9: An illustration of the image partitioning and the structured output word space for maximum margin learning.

vector $\boldsymbol{\alpha}$ such that the prediction error is minimized for all the training instances. That is,

$$\arg\max_{\bar{\mathbf{y}} \in \mathcal{Y}_i} f(\mathbf{x}_i, \bar{\mathbf{y}}) \approx \mathbf{y}_i, \quad i = 1, \cdots, n$$

where $\mathcal{Y}_i = \{\bar{\mathbf{y}} | \sum_{j=1}^{W} \bar{\mathbf{y}}_j = \sum_{j=1}^{W} \mathbf{y}_{ij}\}$. We use $\Phi_i(\bar{\mathbf{y}})$ to denote $\Phi(\mathbf{x}_i, \bar{\mathbf{y}})$. To make the classification be the true output $\mathbf{y}_i$, we must follow

$$\boldsymbol{\alpha}^\top \Phi_i(\mathbf{y}_i) \geq \boldsymbol{\alpha}^\top \Phi_i(\bar{\mathbf{y}}), \quad \forall \bar{\mathbf{y}} \in \mathcal{Y}_i \backslash \{\mathbf{y}_i\}$$

where $\mathcal{Y}_i \backslash \{\mathbf{y}_i\}$ denotes the removal of the element $\mathbf{y}_i$ from the set $\mathcal{Y}_i$. In order to accommodate the classification error on the training examples, we introduce the slack variable ξ_i. The above constraint then becomes

$$\boldsymbol{\alpha}^\top \Phi_i(\mathbf{y}_i) \geq \boldsymbol{\alpha}^\top \Phi_i(\bar{\mathbf{y}}) - \xi_i, \quad \xi_i \geq 0 \quad \forall \bar{\mathbf{y}} \in \mathcal{Y}_i \backslash \{\mathbf{y}_i\}$$

We measure the classification error on the training instances by the loss function, which is the distance between the true output $\mathbf{y}_i$ and the prediction $\bar{\mathbf{y}}$. The loss function measures the goodness of the learning model. The standard zero-one classification loss is not suitable for the structured output space. We define the loss function $l(\bar{\mathbf{y}}, \mathbf{y}_i)$ as the number of the different entries in these two vectors. We include the loss function in the constraints as proposed by Taskar et al. [200]:

$$\boldsymbol{\alpha}^\top \Phi_i(\mathbf{y}_i) \geq \boldsymbol{\alpha}^\top \Phi_i(\bar{\mathbf{y}}) + l(\bar{\mathbf{y}}, \mathbf{y}_i) - \xi_i$$

We interpret $\frac{1}{\|\boldsymbol{\alpha}\|} \boldsymbol{\alpha}^\top [\Phi_i(\mathbf{y}_i) - \Phi_i(\bar{\mathbf{y}})]$ as the margin of $\mathbf{y}_i$ over another $\bar{\mathbf{y}} \in \mathcal{Y}^{(i)}$. We then rewrite the above constraint as $\frac{1}{\|\boldsymbol{\alpha}\|} \boldsymbol{\alpha}^\top [\Phi_i(\mathbf{y}_i) - \Phi_i(\bar{\mathbf{y}})] \geq \frac{1}{\|\boldsymbol{\alpha}\|} [l(\bar{\mathbf{y}}, \mathbf{y}_i) - \xi_i]$. Thus, minimizing $\|\boldsymbol{\alpha}\|$ maximizes such a margin.

The goal now is to solve the optimization problem

$$\min \quad \frac{1}{2} \|\boldsymbol{\alpha}\|^2 + C \sum_{i=1}^{n} \xi_i^r \tag{3.38}$$

$$s.t. \quad \boldsymbol{\alpha}^\top \Phi_i(\mathbf{y}_i) \geq \boldsymbol{\alpha}^\top \Phi_i(\bar{\mathbf{y}}) + l(\bar{\mathbf{y}}, \mathbf{y}_i) - \xi_i$$

$$\forall \bar{\mathbf{y}} \in \mathcal{Y}_i \backslash \{\mathbf{y}_i\}, \quad \xi_i \geq 0, \quad i = 1, \cdots, n$$

where $r = 1, 2$ corresponds to the linear or quadratic slack variable penalty. $C > 0$ is a constant that controls the trade-off between the training error minimization and the margin maximization.

Note that in the above formulation, we do not introduce the relationships between different words in the word space. However, the relationships between different words are implicitly included in the VRep-word pairs because the related words are more likely to occur together. Thus, Equation 3.38 is, in fact, a structured optimization problem.

Maximum margin learning has been studied extensively in the recent machine learning literature due to the important multimedia data mining applications [55, 206, 201, 9]. The challenge of learning with structured output variables is that the number of the structures is exponential in terms of the size of the structure output space. Thus, the problem is intractable if we treat each structure as a separate class. Consequently, the classic multiclass approach is not well fitted into the learning with structured output variables.

As an effective approach to this problem, the maximum margin principle has received substantial attention since it was used in the support vector machine (SVM) [207]. In addition, the perceptron algorithm is also used to explore the maximum margin classification [84]. Taskar et al [200] reduce the number of the constraints by considering the dual of the loss-augmented problem. However, the number of the constraints in their approach is still large for a large structured output space and a large training set.

For learning with structured output variables, Tsochantaridis et al [206] propose a cutting plane algorithm which finds a small set of active constraints. One issue of this algorithm is that it needs to compute the most violated constraint which would involve another optimization problem in the output space. Recently, Guo et al propose the Enhanced Maximum Margin Learning framework (called EMML) [95]. In EMML, instead of selecting the most violated constraint, they arbitrarily select a constraint which violates the optimality condition of the optimization problem. Thus, the selection of the constraints does not involve any optimization problem. Osuna et al [162] propose the decomposition algorithm for the support vector machine. EMML is proposed to extend this idea to the scenario of learning with structured output variables. The rest of this section is to introduce the EMML framework.

One can solve the optimization problem Equation 3.38 in the primal space — the space of the parameters $\boldsymbol{\alpha}$. In fact, this problem is intractable when the structured output space is large because the number of the constraints is exponential in terms of the size of the output space. As in the traditional support vector machine, the solution can be obtained by solving this quadratic optimization problem in the dual space — the space of the Lagrange multipliers. Vapnik [207] and Boyd and Vandenberghe [28] have an excellent review for the related optimization problem.

The dual problem formulation has an important advantage over the primal problem: it only depends on the inner products in the joint feature representation defined by Φ, allowing the use of a kernel function. We introduce the Lagrange multiplier $\mu_{i,\bar{\mathbf{y}}}$ for each constraint to form the Lagrangian. We define $\Phi_{i,\mathbf{y}_i,\bar{\mathbf{y}}} = \Phi_i(\mathbf{y}_i) - \Phi_i(\bar{\mathbf{y}})$ and the kernel function $K((\mathbf{x}_i,\bar{\mathbf{y}}),(\mathbf{x}_j,\tilde{\mathbf{y}})) = \langle \Phi_{i,\mathbf{y}_i,\bar{\mathbf{y}}}, \Phi_{j,\mathbf{y}_j,\tilde{\mathbf{y}}} \rangle$. The derivatives of the Lagrangian over $\boldsymbol{\alpha}$ and ξ_i should be equal to zero. Substituting these conditions into the Lagrangian, we obtain the following Lagrange dual problem

$$\min \frac{1}{2} \sum_{\substack{i,j \\ \bar{\mathbf{y}} \neq \mathbf{y}_i \\ \tilde{\mathbf{y}} \neq \mathbf{y}_j}} \mu_{i,\bar{\mathbf{y}}} \mu_{j,\tilde{\mathbf{y}}} K((\mathbf{x}_i, \bar{\mathbf{y}}), (\mathbf{x}_j, \tilde{\mathbf{y}})) \sum_{\substack{i \\ \bar{\mathbf{y}} \neq \mathbf{y}_i}} \mu_{i,\bar{\mathbf{y}}} l(\bar{\mathbf{y}}, \mathbf{y}_i) \qquad (3.39)$$

$$s.t. \sum_{\bar{\mathbf{y}} \neq \mathbf{y}_i} \mu_{i,\bar{\mathbf{y}}} \leq C \quad \mu_{i,\bar{\mathbf{y}}} \geq 0 \quad i = 1, \cdots, n$$

After this dual problem is solved, we have $\boldsymbol{\alpha} = \sum_{i,\bar{\mathbf{y}}} \mu_{i,\bar{\mathbf{y}}} \Phi_{i,\mathbf{y}_i,\bar{\mathbf{y}}}$.

For each training example, there are a number of constraints related to it. We use the subscript i to represent the part related to the i-th example in the matrix. For example, let $\boldsymbol{\mu}_i$ be the vector with entries $\mu_{i,\bar{\mathbf{y}}}$. We stack the $\boldsymbol{\mu}_i$ together to form the vector $\boldsymbol{\mu}$. That is $\boldsymbol{\mu} = [\boldsymbol{\mu_1}^\top \cdots \boldsymbol{\mu_n}^\top]^\top$. Similarly, let $\mathbf{S}_i$ be the vector with entries $l(\bar{\mathbf{y}}, \mathbf{y}_i)$. We stack $\mathbf{S}_i$ together to form the vector $\mathbf{S}$. That is, $\mathbf{S} = [\mathbf{S}_1^\top \cdots \mathbf{S}_n^\top]^\top$. The lengths of $\boldsymbol{\mu}$ and $\mathbf{S}$ are the same. We define $\mathbf{A}_i$ as the vector which has the same length as that of $\boldsymbol{\mu}$, where $\mathbf{A}_{i,\bar{\mathbf{y}}} = 1$ and $\mathbf{A}_{j,\bar{\mathbf{y}}} = 0$ for $j \neq i$. Let $\mathbf{A} = [\mathbf{A}_1 \cdots \mathbf{A}_n]^\top$. Let matrix $\mathbf{D}$ represent the kernel matrix where each entry is $K((\mathbf{x}_i, \bar{\mathbf{y}}), (\mathbf{x}_j, \tilde{\mathbf{y}}))$. Let $\mathbf{C}$ be the vector where each entry is constant C.

With the above notations, we rewrite the Lagrange dual problem as follows:

$$\min \frac{1}{2} \boldsymbol{\mu}^\top \mathbf{D} \boldsymbol{\mu} - \boldsymbol{\mu}^\top \mathbf{S} \qquad (3.40)$$

$$s.t. \ \mathbf{A}\boldsymbol{\mu} \preceq \mathbf{C}$$

$$\boldsymbol{\mu} \succeq 0$$

where $\preceq$ and $\succeq$ represent the vector comparison defined as entry-wise less than or equal to and greater than or equal to, respectively.

Equation 3.40 has the same number of the constraints as Equation 3.38. However, in Equation 3.40 most of the constraints are lower bound constraints ($\boldsymbol{\mu} \succeq 0$) which define the feasible region. Other than these lower bound constraints, the rest constraints determine the complexity of the optimization problem. Therefore, the number of constraints is considered to be reduced in Equation 3.40. However, the challenge still exists to solve it efficiently since the number of the dual variables is still huge. Osuna et al. [162] propose a decomposition algorithm for the support vector machine learning over large data sets. Guo et al [95] extend this idea to learning with the structured output space with their EMML framework as follows. They decompose the constraints of the optimization problem Equation 3.38 into two sets: the working set B and the nonactive set N. The Lagrange multipliers are also correspondingly partitioned into two parts $\boldsymbol{\mu}_B$ and $\boldsymbol{\mu}_N$. The interest is then in the subproblem defined only for the dual variable set $\boldsymbol{\mu}_B$ when keeping $\boldsymbol{\mu}_N = 0$.

This subproblem is formulated as follows.

$$\min \ \frac{1}{2}\boldsymbol{\mu}^\top \mathbf{D}\boldsymbol{\mu} - \boldsymbol{\mu}^\top \mathbf{S} \tag{3.41}$$

$$s.t. \ \mathbf{A}\boldsymbol{\mu} \preceq \mathbf{C}$$

$$\boldsymbol{\mu}_B \succeq 0, \quad \boldsymbol{\mu}_N = 0$$

It is clearly true that we can move those $\mu_{i,\bar{\mathbf{y}}} = 0, \mu_{i,\bar{\mathbf{y}}} \in \boldsymbol{\mu}_B$ to set $\boldsymbol{\mu}_N$ without changing the objective function. Furthermore, we can move those $\mu_{i,\bar{\mathbf{y}}} \in \boldsymbol{\mu}_N$ satisfying certain conditions to set $\boldsymbol{\mu}_B$ to form a new optimization subproblem which yields a strict decrease in the objective function in Equation 3.40 when the new subproblem is optimized. This property is guaranteed by the following theorem [95].

THEOREM 3.2
Given an optimal solution of the subproblem defined on $\boldsymbol{\mu}_B$ in Equation 3.41, if the following conditions hold true:

$$\exists i, \quad \sum_{\bar{\mathbf{y}}} \mu_{i,\bar{\mathbf{y}}} < C$$

$$\exists \mu_{i,\bar{\mathbf{y}}} \in \boldsymbol{\mu}_N, \quad \boldsymbol{\alpha}^\top \Phi_{i,\mathbf{y}_i,\bar{\mathbf{y}}} - l(\bar{\mathbf{y}}, \mathbf{y}_i) < 0 \tag{3.42}$$

the operation of moving the Lagrange multiplier $\mu_{i,\bar{\mathbf{y}}}$ satisfying Equation 3.42 from set $\boldsymbol{\mu}_N$ to set $\boldsymbol{\mu}_B$ generates a new optimization subproblem that yields a strict decrease in the objective function in Equation 3.40 when the new subproblem in Equation 3.41 is optimized.

In fact, the optimal solution is obtained when there is no Lagrange multiplier satisfying the condition Equation 3.42. This is guaranteed by the following theorem [95].

THEOREM 3.3
The optimal solution of the optimization problem in Equation 3.40 is achieved if and only if the condition Equation 3.42 does not hold true.

The above theorems suggest the EMML algorithm listed in Algorithm 2. The correctness (convergence) of the EMML algorithm is provided by Theorem 3.4 [95].

THEOREM 3.4
The EMML algorithm converges to the global optimal solution in a finite number of iterations.

Note that in Step 5 of Algorithm 2, we only need find one dual variable satisfying Equation 3.42. We need examine all the dual variables in the set

---

Algorithm 2 EMML Algorithm

Input: n labeled examples, dual variable set $\boldsymbol{\mu}$
Output: Optimized $\boldsymbol{\mu}$
Method:

1: Arbitrarily decompose $\boldsymbol{\mu}$ into two sets: $\boldsymbol{\mu_B}$ and $\boldsymbol{\mu_N}$.
2: Solve the subproblem in Equation 3.41 defined by the variables in $\boldsymbol{\mu_B}$.
3: While there exists $\mu_{i,\bar{y}} \in \boldsymbol{\mu_B}$ such that $\mu_{i,\bar{y}} = 0$, move it to set $\boldsymbol{\mu_N}$.
4: While there exists $\mu_{i,\bar{y}} \in \boldsymbol{\mu_N}$ satisfying condition Equation 3.42, move it to set $\boldsymbol{\mu_B}$. If no such $\mu_{i,\bar{y}} \in \boldsymbol{\mu_N}$ exists, the iteration exits.
5: Goto Step 2.

$\boldsymbol{\mu_N}$ only when no dual variable satisfies Equation 3.42. It is fast to examine the dual variables in the set $\boldsymbol{\mu_N}$ even if the number of the dual variables is large. In addition to demonstrating that EMML has a powerful multimodal data mining capability, Guo et al [95] also demonstrate that EMML is about 70 times faster in learning than the method proposed by Taskar et al [200].

3.9   Boosting

Boosting refers to a family of machine learning meta-algorithms for performing supervised learning. The motivation to develop boosting algorithms is based on the question posed by Kearns [123]: can a set of weak learners create a single strong learner? Here a weak learner refers to a classifier that is weakly correlated with the true classification, whereas in contrast, a strong learner is a classifier that is almost always well-correlated with the true classification.

The investigation to Kearns' question has led to the development of the boosting algorithm family as an important contribution to the machine learning and statistics literature. While boosting is not algorithmically constrained, most boosting algorithms follow a template. Typically, boosting occurs in iterations, by incrementally adding weak learners to a final strong learner. At every iteration, a weak learner learns the training data with respect to a distribution. The weak learner is then added to the accumulative strong learner. This is typically performed by weighting the weak learner in a certain way, relating to the weak learner's accuracy. After the weak learner is added to the accumulative strong learner, the data are re-weighted in a reinforced way. For instance, examples that were misclassified now are given weight and examples that were classified correctly now are forced to lose weight (certain boosting algorithms, on the other hand, actually decrease the weight for repeatedly misclassified examples, such as the boost by majority and BrownBoost [82]).

Thus, subsequent weak learners focus more on the examples that previous weak learners have misclassified.

As a family, there are many boosting algorithms proposed in the literature. The original algorithms proposed by Schapire (a recursive majority formulation [186]) and by Freund (boost by majority [81]) were not adaptive and were unable to take full advantage of the weak learners. Also in the literature, several algorithms refer to themselves as "boosting algorithms", and they are quite effective. However, in terms of the *probably approximately correct* learning (PAC) model [132], only provable boosting algorithms may be called "boosting algorithms". Consequently, algorithms that are similar to boosting in spirit, but not PAC-boosters, are sometimes called "leveraging algorithms". These can be quite effective machine learning algorithms [132].

Given the many boosting algorithms, the main variation depends on how to give the weight to the training data points and hypotheses. AdaBoost [83] is a very popular boosting algorithm and is considered as a classic boosting algorithm in the literature; historically, this algorithm may also be considered as the most significant method, as it was the first algorithm developed in the literature that could adapt to the weak learners. Nevertheless, there are many more effective boosting algorithms developed in the literature, such as LPBoost [57], TotalBoost [215], BrownBoost [82], MadaBoost [64], and gradient descent boosting tree [85]. Many boosting algorithms fit into the AnyBoost framework [147], which shows that boosting performs the gradient descent in the function space with a convex cost function. In the rest of this section, we give an introduction to AdaBoost.

AdaBoost, standing for Adaptive Boosting, was developed by Freund and Schapire [83]. It is a meta-algorithm, in the sense that it can be used in conjunction with many other learning algorithms to improve their performance. AdaBoost is adaptive in the sense that subsequently built classifiers are trained in favor of those training samples misclassified by the previous classifiers. On the other hand, AdaBoost is sensitive to noisy data and outliers. Nevertheless, it is less susceptible to the overfitting problem than most learning algorithms in the literature.

Given a series of iterations $t = 1, 2, ..., T$, AdaBoost calls a weak classifier repeatedly. For each call a distribution of weights D_t is updated that reflects the importance of the samples in the training data for the classification. In each iteration, the weight of each incorrectly classified sample is increased (or alternatively, the weight of each correctly classified sample is decreased), such that the new classifier focuses more on those samples. Algorithm 3 lists the AdaBoost algorithm.

In Algorithm 3, after selecting an optimal classifier h_t for the distribution D_t, the training samples $\mathbf{x_i}$ that the classifier h_t has identified correctly are weighted less and those that h_t has identified incorrectly are weighted more. Consequently, when the algorithm uses the classifiers on the distribution D_{t+1}, it selects a classifier that better identifies those training samples that the previous classifiers have missed. Boosting may be considered as a minimization

---

Algorithm 3 AdaBoost Algorithm

Input: $(x_i, y_i), i = 1, \ldots, m$ where $x_i \in X, y_i \in Y = \{-1, +1\}$
Output: A strong classifier $H(.)$
Method:

1: Initialize $t = 1$
2: Initialize $D_t(i) = \frac{1}{m}$ for all $i = 1, \ldots, m$
3: **for** $t = 1$ to T **do**
4:    Train weak learner using distribution D_t
5:    Obtain weak hypothesis $h_t : X \rightarrow \{-1, +1\}$ with error $\epsilon_t = Pr_{i \sim D_t}(h_t(x_i) \neq y_i)$
6:    Choose $\alpha_t = \frac{1}{2}\ln(\frac{1-\epsilon_t}{\epsilon_t})$
7:    Update:

$$D_{t+1}(i) = \frac{D_t(i)}{Z_t} \times \{ \begin{matrix} e^{-\alpha_t} & \text{if } h_t(x_i) = y_i \\ e^{\alpha_t} & \text{if } h_t(x_i) \neq y_i \end{matrix} = \frac{D_t(i)\exp(-\alpha_t y_i h_t(x_i))}{Z_t}$$

   where Z_t is a normalization factor (chosen so that D_{t+1} will be a distribution)
8: **end for**
9: Output the final hypothesis:

$$H(x) = \text{sign}(\sum_{t=1}^{T} \alpha_t h_t(x))$$

---

of a convex loss function over a convex set of functions $\sum_i e^{-y_i f(\mathbf{x}_i)}$. Specifically, the loss being minimized is the exponential loss with the function we expect to obtain as $f = \sum_t \alpha_t h_t$.

3.10 Multiple Instance Learning

In addition to the typical discriminative learning and generative learning methods, multiple instance learning is considered as a new type of learning method recently developed in the machine learning community and is applied in multimedia data mining.

Originally independently proposed by Dietterich et al [59], Auer [11], and Maron and Lozano-Perez [146], *multiple instance learning* addresses a special type of learning problems in which there are ambiguities involved during the training. Consequently, multiple instance learning sometimes is also called learning with ambiguities.

Unlike the classic classification training in which each example instance is given a definite label, in multiple instance learning, a label is not given to an instance; instead, it is given to a group of instances. In the classic multiple instance learning definition [59], training labels are given to groups of example instances where each group has multiple instances which may either be relevant (and thus would be labeled as "yes") to a specific class or be irrelevant (and thus would be labeled as "no") to the specific class. Such a group of instances is called a "bag". In other words, the training data labels are only given to bags of instances instead of instances themselves in the classic training scenario. A bag is labeled "yes" if and only if there is at least one instance in the bag that is relevant, and a bag is labeled "no" if and only if there is no instance in the bag that is relevant.

Ever since the seminal work of multiple instance learning appeared in the machine learning literature, this type of learning approach has been immediately applied to multimedia data mining [225, 237]. This is due to the fact that in many multimedia data mining applications, the training labels are not available to instances, but may be available to groups of instances, i.e., bags of instances. For example, in image data mining, regions or blocks of an image may be considered as instances and the image itself may be considered as a bag. If the image contains a house in the foreground and mountains, sky, and grass in the background, the image can clearly be labeled as "house". But this label is typically given to the whole image even though the actual reference of this label is to one of the regions (or objects) contained in this image — the house, with the other regions or objects irrelevant to this label (i.e., the mountains, sky, and grass). More recent work on applying multiple instance learning to multimedia data mining includes [235, 256]. Chen et al

[46] recently add the embedded instance selection principle into the classic multiple instance learning algorithm, resulting in a better learning performance, and also have applied this method to image data mining. Zhou and Xu [254] have established the connection between multiple instance learning and semi-supervised learning, which is the topic of the next section.

On the other hand, ever since the development of multiple instance learning in the literature, there are several classic multiple instance learning methods proposed in the machine learning community, including the diverse density method [146], the chi-square method [149], and the EM-DD method [236]. Below we introduce the co-learning framework developed based on the multiple instance learning paradigm and implemented using the Diverse Density algorithm in the specific application of image annotation [248].

We first consider the scenario that the whole database is initially used as the training set to build up the database indexing before the database is allowed to evolve under the assumption that the word vocabulary stays the same. We will relax this assumption in the scalability analysis in Section 3.10.6.

In the rest of this section, we use calligraphic letters to denote the set variables or functions, and to use regular letters to denote regular variables or functions. A database $\mathcal{D} = \{\mathcal{I}, \mathcal{W}\}$ consists of two parts, an image collection $\mathcal{I}$ and a vocabulary collection $\mathcal{W}$. A collection of images $\mathcal{I} = \{I_i, i = 1, ..., N\}$ is the whole database images used as the training set; $N = |\mathcal{I}|$; for each image I_i, there is a set of words annotating this image $\mathcal{W}_i = \{w_{ij}, j = 1, ..., N_i\}$; the whole vocabulary set of the database is $\mathcal{W}$; and $M = |\mathcal{W}| = |\bigcup_{i=1}^{N} \mathcal{W}_i|$. We define a block as a subimage of an image such that the image is partitioned into a set of blocks and all the blocks of this image share the same resolution. We define a VRep (visual representative) as a representative of a set of all the blocks for all the images in the database that appear visually similar to each other. A VRep of an image may be represented as a feature vector in a feature space.

Before we present the co-learning framework, we first make a few assumptions.

- A semantic concept corresponds to one or multiple words, and a word corresponds to one semantic concept for each image. Consequently, semantic concepts may be represented in words.

- A semantic concept corresponds to one or multiple VReps, and a VRep corresponds to one or multiple semantic concepts.

- A word corresponds to one or multiple VReps, and a VRep corresponds to one or multiple words.

- An image may have one or more words for annotation.

3.10.1   Establish the Mapping between the Word Space and the Image-VRep Space

For each image I_i, we partition it into a set of exclusive blocks B_{ij}, i.e.,

$$I_i = \bigcup_j B_{ij}, j = 1, ..., n_i \qquad B_{ij} \cap B_{ih} = \emptyset, j \neq h \qquad (3.43)$$

where n_i is a function of the resolution of I_i such that the resolution of B_{ij} is no less than a threshold. If all the images in the database are in the same resolution, all the n_i's are the same as a constant. Since each block may be represented as a feature vector in a feature space, for all the blocks of all the images in the database, a nearest neighbor clustering in the feature space leads to a partition of the whole block feature vectors in the feature space into a finite number of clusters such that each cluster is represented by its centroid; let L be the number of such clusters. This centroid is a VRep corresponding to this cluster for all the images in the database. Consequently, the whole VRep set in the database is

$$\mathcal{V} = \{v_i | i = 1, ..., L\} \qquad (3.44)$$

Thus, each image I_i may be represented by a subset of $\mathcal{V}$. Each VRep is represented as a feature vector in the feature space and corresponds to a subset of all the images in the database such that this VRep appears in those images in the subset; i.e., for each VRep v_i, there is a subset $\mathcal{I}_{v_i}$ of the images in the database such that

$$\mathcal{I}_{v_i} = \{I_h | h = 1, ..., n_{v_i}\} \qquad (3.45)$$

where $n_{v_i} = |\mathcal{I}_{v_i}|$.

Once we have obtained all the VReps for the images in the database, we sort all the textual vocabulary words in W (say alphabetically), and for each word w_k, there is a corresponding set of images, $\mathcal{S}_k$, such that this word w_k appears in the annotation of each of the images in the set. Since each image is represented as a set of decomposed blocks, $\mathcal{S}_k$ may be represented as

$$\mathcal{S}_k = \{I_{k_i} | I_{k_i} = \bigcup_j B_{k_{ij}}, j = 1, ..., n_{k_i}\} \qquad (3.46)$$

where $B_{k_{ij}}$ is the jth block in image I_{k_i}. For each block $B_{k_{ij}}$ in image I_{k_i}, a feature vector $f_{k_{ij}}$ in the feature space is used to represent the block. In order to establish the relationship between the word space and the image-VRep space, we map the problem to a multiple instance learning problem [59]. A general multiple instance learning problem is to learn a function $y = F(x)$, where we are given multiple samples of x represented as bags, and each bag has ambiguities represented by the multiple instances of x. Here the problem is that each bag is an image, and all the instances of this bag are the blocks represented by the corresponding feature vectors; the y here is a word vector instead of a value in the range of $[0, 1]$ in the classic version of multiple instance learning, consisting of all the words given in the training set that correspond to a specific VRep; the function to be learned each time is the function of a VRep mapping to the words. Specifically, for each word $w_k \in \mathcal{W}$, we use multiple instance learning to apply to the whole image database to obtain the optimal block feature vector t_k. Given the distribution of all the $f_{k_{ij}}$ corresponding to the image set $\mathcal{S}_k$ in the feature space, using the diverse density algorithm of multiple instance learning [146], we are able to immediately obtain the optimal block feature vector t_k:

$$t_k = \arg\max_t \prod_t P(t \in I | I \in \mathcal{S}_k) \prod_t P(t \in I | I \notin \mathcal{S}_k) \qquad (3.47)$$

where $P(.|.)$ is a posterior probability. Now we have established the one-to-one mapping between the word w_k and the block feature vector t_k. Then we use the nearest neighbor clustering to identify all the closest VReps v_{k_l} such that

$$\|t_k - v_{k_l}\| < T_k \qquad (3.48)$$

where T_k is a threshold. Denote the set of those VReps that satisfy this constraint as $\mathcal{V}_k$,

$$\mathcal{V}_k = \{v_{k_l} | l = 1, ..., n_{w_k}\} \qquad (3.49)$$

where n_{w_k} is the number of such VReps satisfying this constraint. Thus, for each word w_k, there is a corresponding set of VReps $\mathcal{V}_k$ that are close to t_k subject to the threshold T_k. In addition, according to Equation 3.45, each such VRep v_{k_l} has an associated image set $\mathcal{I}_{k_l}$ such that all the images in the set have this VRep. For each such image $I_{k_{l_i}} \in \mathcal{I}_{k_l}$, using the mixture of Gaussian model [60], we compute the posterior probability $P(I_{k_{l_i}} | w_k)$. Then, we rank all the images in the set $\mathcal{I}_{k_l}$ by the posterior probability $P(I_{k_{l_i}} | w_k)$. We denote such a ranked list of images in the database as $\mathcal{L}_k$. Hence, for each word w_k, there is a corresponding ranked list of images in the database

$$\mathcal{L}_k = \{I_{k_h} | h = 1, ..., |\mathcal{L}_k|\} \qquad (3.50)$$

i.e.,

$$w_k \leftrightarrow \mathcal{L}_k \qquad (3.51)$$

Similarly, we use multiple instance learning to learn the function $y = F'(x)$ where x's are the ambiguous instances of the annotation words for an image and y is the set of the corresponding VReps to a word; here again the bag is an image. Specifically, for each VRep v_i, according to Equation 3.45, there is a corresponding image set $\mathcal{I}_{v_i}$, and for each image $I_{v_{ij}} \in \mathcal{I}_{v_i}$, there is a corresponding annotation word set $\mathcal{W}_{v_{ij}}$:

$$\mathcal{W}_{v_{ij}} = \{w_{v_{ij}}^h | h = 1, ..., |\mathcal{W}_{v_{ij}}|\} \qquad (3.52)$$

Thus, using the diverse density algorithm of multiple instance learning [146] again, we are able to obtain the optimal annotation word w_k corresponding to the image set $\mathcal{I}_{v_i}$:

$$w_k = \arg\max_w \prod_w P(w \in \mathcal{W}_{v_{ij}} | I \in \mathcal{I}_{v_i})$$

$$\times \prod_w P(w \in \mathcal{W}_{v_{ij}} | I \notin \mathcal{I}_{v_i}) \qquad (3.53)$$

Similarly, we may use the same algorithm to compute the ith best annotation word corresponding to VRep v_i. Consequently, for every VRep v_i, there is a corresponding ranked list of annotation words $\mathcal{L}_{v_i}$, i.e.,

$$v_i \leftrightarrow \mathcal{L}_{v_i} \qquad (3.54)$$

Finally, for every VRep $v_i \in \mathcal{V}$, we compute the prior probability $P(v_i)$ by determining the relative occurrence frequency of v_i in the whole image database. Similarly, for every word $w_k \in \mathcal{W}$, we compute the prior probability $P(w_k)$ by determining the relative occurrence frequency of w_k for all the images in the database.

Given this learned correspondence relationship between the word space and the imagery space, we have completed the co-learning for indexing the database as part of the framework. Now we are ready for across-modality retrieval and mining.

3.10.2 Word-to-Image Querying

If a query is given by several words for mining and retrieving images from the database, we assume that the query consists of words $w_{q_k}, k = 1, ..., p$. We also assume that all the query words are within the textual vocabulary of the training data. Since each w_{q_k} has a corresponding ranked list of images $\mathcal{L}_k$, we just need to merge these p ranked lists $\mathcal{L}_k$, $k = 1, ..., p$ by $P(I_{k_i} | w_{q_k})$ for all the different images I_{k_i}.

Since the bottleneck of the computation is on merging the p ranked lists, the total computation is in $O(p|\mathcal{L}_k|)$, which is independent of the database scale $O(M, N)$. Hence, this querying complexity is $O(1)$.

3.10.3 Image-to-Image Querying

If a query is given by several images for mining and retrieving images from the database, we assume that the query consists of images $I_{q_k}, k = 1, ..., p$. These images may or may not be necessarily from the database; however, we assume that these images follow the same feature distributions as those in the database. For each query image I_{q_k}, we partition it into p_k blocks following the definition in Equation 3.43, and extract the feature vector $f_{q_{kl}}$ for each block $B_{q_{kl}}$. For each $f_{q_{kl}}$, we compute the similarity distances to all the VReps

v_i in the feature space. Based on the similarity distances and the assumption that the features in the query images follow the same distributions as those of the images in the database, each $B_{q_{kl}}$ is replaced with the corresponding closest VRep v_i in the feature space. From Equation 3.45, each v_i has a corresponding image set $\mathfrak{I}_{v_i}$; we assume that there are in total r_k such VReps v_i found in the query image I_{q_k} and $r_k \leq p_k$. Let $\mathcal{S}_{q_k}$ be the largest common set of the r_k image sets $\mathfrak{I}_{v_i}$.

On the other hand, for each VRep v_i of I_{q_k}, we immediately have a ranked word list $\mathcal{U}_{v_i}$ based on $P(w_k|v_i)$. We merge the r_k ranked lists based on $P(w_k|v_i)P(v_i|I_{q_k})$ to form a new ranked list $\mathcal{U}_{q_k}$, where $P(v_i|I_{q_k})$ is the occurrence frequency of the VRep v_i appearing in the image I_{q_k}. For all the words in the list $\mathcal{U}_{q_k}$ (in the implementation we may truncate to the top few words for the list), we use the word-to-image querying scheme in Section 3.10.2 to generate a ranked image list $\mathcal{L}_{q_k}$. $\mathcal{L}_{q_k}$ is then further trimmed such that only those images that are in $\mathcal{S}_{q_k}$ survive with the same relative ranking order in $\mathcal{L}_{q_k}$. Finally, we merge the p ranked lists $\mathcal{L}_{q_k}, k = 1, ..., p$.

Given an appropriate hashing function for all the images in the database, this querying may be done in $O(p|\mathcal{U}_{q_k}|)$, which again is independent of the database scale $O(M, N)$. Hence, this querying complexity is $O(1)$.

3.10.4   Image-to-Word Querying

If the query is given by several images for word mining and retrieval, i.e., for automatic annotation, we assume that the query consists of p images, $I_{q_k}, k = 1, ..., p$. Similar to the image-to-image mining and querying in Section 3.10.3, each query image I_{q_k} is decomposed into several VReps, and assume that the p query images have in total s_k VReps $v_i, i = 1, ..., s_k$. Let $P(v_i|I_q)$ be the relative frequency of the VRep v_i in all the query images $I_{q_k}, k = 1, ..., p$. Since each VRep v_i has a corresponding ranked list of words $\mathcal{U}_{v_i}$ based on $P(w_k|v_i)$, the final mining and retrieval are the merged ranked list of words based on $P(w_k|v_i)P(v_i|I_q)$ from the s_k ranked lists $\mathcal{U}_{v_i}$.

Similarly, this querying is done in $O(s_k|\mathcal{U}_{v_i}|)$, which is again independent of the database scale $O(M, N)$. Hence, this querying complexity is $O(1)$.

3.10.5   Multimodal Querying

If the query is given by a combination of a series of words and a series of images for multimedia data mining and retrieval, without loss of generality, we perform multimodal image querying as follows. We use the word-to-image querying in Section 3.10.2 and the image-to-image querying in Section 3.10.3, respectively, and finally merge the queryings together based on their corresponding posterior probabilities.

Clearly, this querying is in $O(\max\{p|\mathcal{L}_k|, p|\mathcal{U}_{q_k}|, s_k|\mathcal{U}_{v_i}|\}) = O(1)$, independent of the database scale $O(M, N)$. Consequently, the general multimodal querying complexity is again $O(1)$.

3.10.6   Scalability Analysis

As is given in the analysis presented in Sections. 3.10.2 and 3.10.5, it is clear that it only takes a constant time to process any type of query for the mining and querying. At the same time, the mining effectiveness is also independent of the database scale. This advantage is supported and verified in the empirical evaluations [248]. Thus, this co-learning framework is highly scalable. Note that many existing multimodal data mining methods in the literature have a complexity dependent upon the databases' scales, typically linear, and their mining effectiveness degrades substantially when the databases scale up. With this advantage, the co-learning framework substantially advances the literature and excels these existing methods in the sense that it pushes to a new horizon a major step forward toward a real-world application with a very large-scale database.

3.10.7   Adaptability Analysis

In this section, we show through the complexity analysis that the co-learning framework is adaptable very well when the database undergoes dynamic changes. This analysis is also consistent with the empirical evaluations [248]. Specifically, we show that the database indexing updating only takes $O(1)$ computation for the following three exhaustive cases. The first two cases consider the scenario when images are added into or deleted from the database while the accompanying annotation words stay with the same vocabulary of the database; the last case considers the scenario when the vocabulary also changes. For all the cases, we assume that any newly added images follow the same feature distributions as those already in the database.

3.10.7.1   Case 1: When a New Image Is Added into the Database

Let I^{new} be the new image. We first consider the new image added with no annotation. We show that the following steps complete updating the indexing of the database and that in each of the steps only a local change is necessary; i.e., the updating complexity is a constant w.r.t. the database size (N, M).

- **Step 1: Determining VReps**

  Following the definition in Equation 3.43, we partition the image into blocks; based on the assumption that any newly added images follow the same distributions in the feature space as those of the images in the database, from the VRep definitions in the feature space of the images in the database in Equation 3.44, each block of I^{new} is replaced with the nearest VRep in the feature space. This step takes $O(L)$ time. Since $L \ll N, M$, this step is $O(1)$.

- **Step 2: Updating VRep-to-image mapping**

For each VRep v_i of I^{new}, we revise the corresponding image set $\mathcal{I}_{v_i}$ in Equation 3.45 by adding I^{new} into the set $\mathcal{I}_{v_i}$ to become $\mathcal{I}_{v_i}{}^{new}$ and thus revise the occurrence frequency as the prior probability $P(v_i)$ (through incrementing both the numerator and the denominator of the previous prior probability). We assume that the list $\mathcal{I}_{v_i}$ is indexed by an array, such that the insertion of a new image takes a constant time. Thus, this step is $O(1)$.

- **Step 3: Determining annotation words**

  In order to determine the annotation words for I^{new}, i.e., to determine the image-to-word mapping for I^{new}, since each VRep in I^{new} has a corresponding ranked list of words in the original database indexing already, the annotation words for I^{new} is the merged ranked list of all the ranked lists of words corresponding to the VReps in I^{new} based on $P(w_k|v_i)P(v_i|I^{new})$ where $P(v_i|I^{new})$ is the relative occurrence frequency of VRep v_i in I^{new}. Let $\mathcal{A}_{I^{new}}$ be this merged ranked list for I^{new}. In practice, this list may be truncated to a top few words as the appropriate annotation words for I^{new} based on the ranking weight $P(w_k|v_i)P(v_i|I^{new})$. The total time is $O(L)$. Since $L \ll N, M$, this step is $O(1)$.

- **Step 4: Updating word-to-VRep mapping**

  For updating the word-to-VRep mapping in the database indexing, in order to avoid redoing the indexing from scratch, we approximate the original mapping in the database indexing from a word to a set of VReps in Equation 3.49 using the weighted VReps instead of the actual block feature vectors of all the images in the feature space in the original database. Thus, we just need to revise the weights (i.e., the occurrence frequencies) of those VReps that appear in I^{new}. From Equation 3.54, each VRep v_i appearing in I^{new} has a corresponding ranked list of annotation words $\mathcal{L}_{v_i}$. Given all the VReps appearing in I^{new}, we immediately have a merged ranked list of annotation words from all the ranked lists $\mathcal{L}_{v_i}$. Since in practice we truncate each ranked word list $\mathcal{L}_{v_i}$ to the top few words, these truncated ranked lists are merged together. Let $\mathcal{C}^t$ be such a merged list. Consequently, we only need to update the word-to-VRep mapping for those words in $\mathcal{C}^t$. Specifically, for each $w_k \in \mathcal{C}^t$, we check the corresponding VRep set $\mathcal{V}_k$ in Equation 3.49 and increment the frequency counts by one for those VReps in $\mathcal{V}_k$ that appear in I^{new}; we then update the optimal centroid $t_k{}^{new}$ based on the updated VRep frequencies in $\mathcal{V}_k$ in the feature space. Finally, we use the same threshold T_k to update the new nearest neighborhood $\mathcal{V}_k{}^{new}$ of $t_k{}^{new}$ based on Equation 3.48. Thus, the complexity of this step is $O(|\mathcal{C}^t|L)$. Since $|\mathcal{C}^t| \ll M$, and $L \ll M, N$, this step is $O(1)$.

- **Step 5: Updating word-to-image mapping**

In order to update the word-to-image mapping list in Equation 3.50 and the prior probability $P(w_k)$, we only need to focus on those words $w_k \in \mathcal{A}_{I^{new}}$ obtained in Step 3. For each word $w_k \in \mathcal{A}_{I^{new}}$, we have the updated corresponding VRep set $\mathcal{V}_k^{new}$ obtained in Step 4. We only need to check those VReps $v_i \in \mathcal{V}_k^{new}$ that were updated for the corresponding image set $\mathcal{I}_{v_i}^{new}$ in Step 2. Let L^u be the number of such VReps in $\mathcal{V}_k^{new}$. Thus, the posterior probability $P(I^{new}|w_k)$ is approximated as

$$P(I^{new}|w_k) = \frac{L^u}{L} \tag{3.55}$$

and the prior probability $P(w_k)$ is approximated as

$$P(w_k) = \frac{L^u}{N} \tag{3.56}$$

Finally, the word-to-image mapping ranked list $\mathcal{L}_k$ in Equation 3.50 is updated by inserting I^{new} based on the weight $P(I^{new}|w_k)$ determined in Equation 3.55. Clearly, this step takes $O(L)$ which is $O(1)$ due to $L \ll M, N$.

When the new image is added with annotation words, $\mathcal{A}_{I^{new}}$ in Step 3 is given and thus Step 3 is skipped; Step 4 now may be only focused on determining the word-to-VRep mapping for those annotation words given in the new image, which can be similarly shown to be done in $O(1)$ time; the rest of the procedure is exactly the same. Consequently, the constant updating conclusion still holds true.

3.10.7.2 Case 2: When an Existing Image Is Deleted from the Database

Let I^{del} be the image in the original database that needs to be deleted. Let $v^i, i = 1, ..., r_d$ be the VReps of I^{del}. Let $w^k, k = 1, ..., s_d$ be the annotation words accompanying I^{del}. We show that the following steps complete the indexing updating with a constant time.

- **Step 1: Updating VRep-to-image mapping**

  For each $v^i, i = 1, ..., r_d$, we remove I^{del} in the corresponding image list $\mathcal{I}_{v^i}$ in Equation 3.45. We then update the prior probability $P(v^i)$ by decrementing the numerator (the occurrence frequency of images with v^i) and the denominator (i.e., N). We assume that the list $\mathcal{I}_{v^i}$ is indexed by an array, such that deleting an image from the list takes a constant time. Thus, this step is $O(1)$.

- **Step 2: Updating word-to-VRep mapping**

  Similar to Step 4 in Case 1, we update the word-to-VRep mapping. Here, instead of incrementing the occurrence frequencies of those v^i

in the feature space, we decrement the occurrence frequencies of the v^i. In practice, it is sufficient to focus on updating the mapping for $w^k, k = 1, ..., s_d$. Since $s_d \ll M$, as is the complexity of Step 4 in Case 1, this step is $O(1)$.

- **Step 3: Updating word-to-image mapping**

  To update the word-to-image mapping, according to Equation 3.50, we just need to remove I^{del} from the ranked list $\mathcal{L}_k$ for each word w^k. The prior probability $P(w^k)$ can also be updated immediately by decrementing the occurrence frequency of w^k. Assuming that the ranked list $\mathcal{L}_k$ is indexed in an array, the removal of an image from this list is a constant operation. Thus, this step is $O(1)$.

3.10.7.3   Case 3: When the Database Vocabulary Changes

Since we only consider the scenario in which text vocabulary in a database is the accompanying information for the annotation of the images in the database, there are only two subcases for the dynamic change of the text vocabulary in the database: (1) when a new image is added into the database with textual annotations using new text vocabulary; and (2) when an existing text word is deleted from the text vocabulary in the database at the time when its accompanying image is deleted from the database. Clearly, Subcase 2 is a special case of Section 3.10.7.2. Thus, we only need to discuss Subcase 1 here.

Let I^{new} be the new image to be added into the database. Let $w^l, l = 1, ..., r_l$ be the annotation words of I^{new} that are from the existing vocabulary in the database, and let $w^k, k = 1, ..., r_k$ be the annotation words new to the database text vocabulary. We show that following a procedure similar to that in Section 3.10.7.1 the database indexing may be updated with a constant time. Specifically, Steps 1, 2, and 5 are exactly the same as the corresponding steps in Section 3.10.7.1, and Step 3 is skipped as the annotation words for I^{new} are given. Thus, we just need to show the remaining **Step 4: updating word-to-VRep mapping** as follows.

For each w^l, the updating is exactly the same as that in Step 4 of Section 3.10.7.1. For each w^k, since all the VReps corresponding to w^k in the feature space are those appearing in I^{new}, following the same procedure in Section 3.10.1 with the same constraint in Equation 3.48 and the threshold T_k, we obtain the optimal VRep feature vector t_k as well as its neighborhood, from which we immediately obtain a list of the VReps corresponding to w^k defined similarly to $\mathcal{V}_k$ in Equation 3.49. Let $\mathcal{V}^k$ be such a VRep list corresponding to $w^k, k = 1, ..., r_k$. Since the number of the VReps appearing in I^{new} is finite, and since $r_k \ll M$, the complexity of this step is $O(1)$.

In summary, we have shown that regardless of how the database changes, the updating of the indexing may be done incrementally with a constant time without needing to redo the indexing from the scratch. This allows that the database indexing can always be updated in a timely manner with an incremental change of the database content. Therefore, this co-learning framework is highly adaptable. This adaptability advantage of the framework enables it to excel many of the peer methods in the literature that do not adapt at all; i.e., the database must be reindexed (or retrained) from scratch even if the database is only incrementally updated. The evaluations of the effectiveness and the superiority of this co-learning framework to a peer, state-of-the-art multimedia data mining method in the literature as well as the co-learning framework's high scalability and adaptability may be found at [248].

3.11   Semi-Supervised Learning

In many multimedia data mining applications, it is common that we do not have the luxury to have plenty of training samples. There are two reasons contributing to the presence of this constraint. First, obtaining the training samples with known labels is typically expensive. Second, in many applications the complexity to obtain the training samples with known labels is so high that we are unable to afford to have many such training samples. Consequently, it is important to live with the reality with only a few training samples in order to still be able to mine the multimedia data. In machine learning literature, the paradigm of semi-supervised learning nicely satisfies this constraint, making it very popular recently in multimedia data mining applications.

Unlike the classic supervised learning in which there is only labeled training data available for training, semi-supervised learning addresses the scenario in which the training data consists of two parts: labeled training data and unlabeled training data. Typically, the labeled training data only constitute a small set, while the unlabeled training data may be a large set, as the latter are easier to obtain in many applications than the first. The key to a successful semi-supervised learning method is to exploit the existence of the large number of unlabeled data samples to help improve the performance of the classifier that would only be trained by the small set of the labeled training data samples.

Ever since the seminal work of co-training [26], semi-supervised learning has been developed as a hot topic in the machine learning literature. Zhu gives a comprehensive review on the early work of semi-supervised learning [257]. Recently, in the context of multimedia data mining, Yao and Zhang have studied the accuracy issue of semi-supervised learning. Specifically, they have addressed how to achieve an optimal accuracy [228] and how to ensure that the accuracy increases with the iterations in semi-supervised learning [227]. In this section, we introduce a recently developed semi-supervised learning framework — a semiparametric regularization based approach [96], which attempts to discover the marginal distribution of the data to learn the parametric function through exploiting the geometric distribution of the data. This learned parametric function can then be incorporated into the supervised learning on the available labeled data as the prior knowledge.

Most of the semi-supervised learning models are based on the *cluster assumption*, which states that the decision boundary should not cross the high-density regions, but instead lie in the low-density regions. In other words, similar data points should have the same label and dissimilar data points should have different labels. The approach introduced here is also based on the cluster assumption. Moreover, we believe that the marginal distribution of the data is determined by the unlabeled examples if there is a small labeled data set available along with a relatively large unlabeled data set, which is the case for many applications. The geometry of the marginal distribution must be considered such that the learned classification or regression function adapts to the data distribution. An example is shown in Figure 3.10 for a binary classification problem. In Figure 3.10(a) the decision function is learned only from the labeled data and the unlabeled data are not used at all. Since the labeled data set is very small, the decision function learned cannot reflect the overall distribution of the data. On the other hand, the marginal distribution of the data described by the unlabeled data has a particular geometric structure. Incorporating this geometric structure into the learning process results in a better classification function, as shown in Figure 3.10(b).

The above observation suggests that the unlabeled data help change the decision function toward the desired direction. Therefore, the question we set for ourselves here is the following:

How do we incorporate the geometric structure of the marginal distribution of the data into the learning such that the resulting decision function $\bar{f}$ reflects the distribution of the data?

A variety of graph based methods is proposed in the literature to achieve this goal. The approach presented here exploits the geometric structure in a different way. This is achieved by a 2-step learning process. The first step is to obtain a parametric function from the unlabeled data which describes the geometric structure of the marginal distribution. This parametric function is obtained by applying the Kernel Principal Component Analysis (KPCA) algorithm to the whole data, including the labeled and unlabeled data. In KPCA, the function to extract the most important principal component is a linear combination of the kernel functions in the Reproducing Kernel Hilbert Space (RKHS), $f(\mathbf{x}) = K(\mathbf{x}, .)\boldsymbol{\alpha}$, where K is a kernel function and $\boldsymbol{\alpha}$ is the coefficients vector. This learned parametric function can be shown to reflect the geometric structure of the marginal distribution of the data. The second step is a supervised learning on the labeled data. To incorporate this parametric function into the supervised learning, we extend the original RKHS to be used in the supervised learning by including this parametric

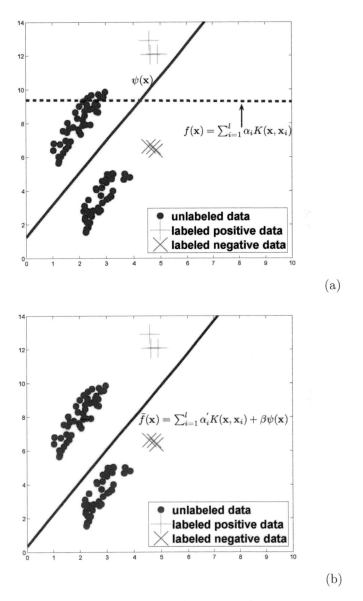

FIGURE 3.10: (a) The decision function (dashed line) learned only from the labeled data. (b) The decision function (solid line) learned after the unlabeled data are considered also.

function learned from the whole data. Consequently, we call this approach a *semiparametric regularization* based semi-supervised learning.

By selecting different loss functions for the supervised learning, we obtain different semi-supervised learning algorithms. We primarily focus on two families of the algorithms: the semiparametric regularized Least Squares (hereafter SpRLS) and the semiparametric regularized Support Vector Machines (hereafter SpSVM). These algorithms demonstrate the state-of-the-art performance on a variety of multimedia data mining tasks.

3.11.1   Supervised Learning

We begin with a brief review of supervised learning. Suppose that there is a probability distribution P on $\mathcal{X} \times \mathcal{Y}, \mathcal{X} \subset \mathbb{R}^n$ according to which data are generated. We assume that the given data consist of l labeled data points $(\mathbf{x}_i, y_i), 1 \leq i \leq l$ which are generated according to P. In this section, we assume the binary classification problem where the labels $y_i, 1 \leq i \leq l$, are binary, i.e., $y_i = \pm 1$.

In the supervised learning scenario, the goal is to learn a function f to minimize the expected loss called risk functional

$$R(f) = \int L(\mathbf{x}, y, f(\mathbf{x})) dP(\mathbf{x}, y) \qquad (3.57)$$

where L is a loss function. A variety of loss functions have been considered in the literature. The simplest loss function is 0/1 loss:

$$L(\mathbf{x}_i, y_i, f(\mathbf{x}_i)) = \begin{cases} 0 & \text{if } y_i = f(\mathbf{x}_i) \\ 1 & \text{if } y_i \neq f(\mathbf{x}_i) \end{cases} \qquad (3.58)$$

In Regularized Least Square (RLS), the loss function is given by

$$L(\mathbf{x}_i, y_i, f(\mathbf{x}_i)) = (y_i - f(\mathbf{x}_i))^2$$

In SVM, the loss function is given by

$$L(\mathbf{x}_i, y_i, f(\mathbf{x}_i)) = \max(0, 1 - y_i f(\mathbf{x}_i))$$

For the loss function Equation 3.58, Equation 3.57 determines the probability of a classification error for any decision function f. In most applications the probability distribution P is unknown. The problem, therefore, is to minimize the risk functional when the probability distribution function $P(\mathbf{x}, y)$ is unknown but the labeled data $(\mathbf{x}_i, y_i), 1 \leq i \leq l$ are given. Thus, we need to consider the empirical estimate of the risk functional [209]:

$$R_{emp}(f) = C \sum_{i=1}^{l} L(\mathbf{x}_i, y_i, f(\mathbf{x}_i)) \qquad (3.59)$$

Table 3.2: Most frequently used kernel functions.

Kernel Name	Kernel Function
polynomial kernel	$K(\mathbf{x}, \mathbf{x}_i) = (\langle \mathbf{x}, \mathbf{x}_i \rangle + c)^d$
Gaussian radial basis function kernel	$K(\mathbf{x}, \mathbf{x}_i) = \exp(-\frac{\|\mathbf{x} - \mathbf{x}_i\|^2}{2\sigma^2})$
sigmoid kernel	$K(\mathbf{x}, \mathbf{x}_i) = \tanh(\kappa \langle \mathbf{x}, \mathbf{x}_i \rangle + \vartheta)$

where $C > 0$ is a constant. We often use $C = \frac{1}{l}$. Minimizing the empirical risk Equation 3.59 may lead to numerical instabilities and bad generalization performance [187]. A possible way to avoid this problem is to add a stabilization (regularization) term $\Theta(f)$ to the empirical risk functional. This leads to better conditioning of the problem. Thus, we consider the following regularized risk functional:

$$R_{reg}(f) = R_{emp}(f) + \gamma \Theta(f)$$

where $\gamma > 0$ is the regularization parameter which specifies the tradeoff between minimization of $R_{emp}(f)$ and the smoothness or simplicity enforced by small $\Theta(f)$. A choice of $\Theta(f)$ is the norm of the RKHS representation of the feature space:

$$\Theta(f) = \|f\|_K^2$$

where $\|.\|_K$ is the norm in the RKHS $\mathcal{H}_K$ associated with the kernel K. Therefore, the goal is to learn the function f which minimizes the regularized risk functional

$$f^* = \arg \min_{f \in \mathcal{H}_K} C \sum_{i=1}^{l} L(\mathbf{x}_i, y_i, f(\mathbf{x}_i)) + \gamma \|f\|_K^2 \qquad (3.60)$$

The solution to Equation 3.60 is determined by the loss function L and the kernel K. A variety of kernels has been considered in the literature. The three most often-used kernel functions are listed in the Table 3.2, where $\sigma > 0, \kappa > 0, \vartheta < 0$.

The following classic Representer Theorem [187] states that the solution to the minimization problem Equation 3.60 exists in $\mathcal{H}_K$ and gives the explicit form of a minimizer.

THEOREM 3.5
Denote by $\Omega : [0, \infty) \to \mathbb{R}$ a strictly monotonic increasing function, by $\mathcal{X}$ a set, and by $\Lambda : (\mathcal{X} \times \mathbb{R}^2)^l \to \mathbb{R} \cup \{\infty\}$ an arbitrary loss function. Then each minimizer $f \in \mathcal{H}_K$ of the regularized risk

$$\Lambda((\mathbf{x}_1, y_1, f(\mathbf{x}_1)), \cdots, (\mathbf{x}_l, y_l, f(\mathbf{x}_l))) + \Omega(\|f\|_K)$$

admits a representation of the form

$$f(\mathbf{x}) = \sum_{i=1}^{l} \alpha_i K(\mathbf{x}_i, \mathbf{x}) \tag{3.61}$$

with $\alpha_i \in \mathbb{R}$.

According to Theorem 3.5, we can use any regularizer in addition to $\gamma\|f\|_K^2$ that is a strictly monotonic increasing function of $\|f\|_K$. This allows us in principle to design different algorithms. Here we take the simplest approach to use the regularizer $\Omega(\|f\|_K) = \gamma\|f\|_K^2$. Given the loss function L and the kernel K, we substitute Equation 3.61 into Equation 3.60 to obtain a minimization problem of the variables $\alpha_i, 1 \leq i \leq l$. The decision function f^* is immediately obtained from the solution to this minimization problem.

3.11.2   Semi-Supervised Learning

In the semi-supervised learning scenario, in addition to l labeled data points $(\mathbf{x}_i, y_i), 1 \leq i \leq l$, we are given u unlabeled data points $\mathbf{x}_i, l + 1 \leq i \leq l + u$, which are drawn according to the marginal distribution $P_{\mathcal{X}}$ of P. The decision function is learned from both the labeled data and the unlabeled data. The semi-supervised learning attempts to incorporate the unlabeled data into the supervised learning in different ways. We here present a semi-supervised learning approach based on *semiparametric regularization* which extends the original RKHS by including a parametric function learned from the whole data, including the labeled and unlabeled data.

In the supervised learning, we may have additional prior knowledge about the solution in many applications. In particular, we may know that a specific parametric component is very likely to be a part of the solution. Or we might want to correct the data for some (e.g., linear) trends to avoid overfitting. Overfitting degrades the generalization performance when outliers exist.

Suppose that this additional prior knowledge is described as a family of parametric functions $\{\psi_p\}_{p=1}^{M} : \mathcal{X} \to \mathbb{R}$. These parametric functions may be incorporated into the supervised learning in different ways. Here we consider the following regularized risk functional

$$\bar{f}^* = \arg\min_{\bar{f}} C \sum_{i=1}^{l} L(\mathbf{x}_i, y_i, \bar{f}(\mathbf{x}_i)) + \gamma\|f\|_K^2 \tag{3.62}$$

where $\bar{f} := f + h$ with $f \in \mathcal{H}_K$ and $h \in span\{\psi_p\}$. Consequently, we extend the original RKHS $\mathcal{H}_K$ by including a family of parametric function ψ_p without changing the norm. The semiparametric representer theorem [187] tells us the explicit form of the solution to Equation 3.62. The following semiparametric representer theorem is an immediate extension of Theorem 3.5.

THEOREM 3.6

Suppose that in addition to the assumptions of Theorem 3.5 we are given a set of M real valued functions $\{\psi_p\}_{p=1}^{M} : \mathcal{X} \to \mathbb{R}$, with the property that the $l \times M$ matrix $(\psi_p(\mathbf{x}_i))_{ip}$ has rank M. Then for any $\bar{f} := f + h$ with $f \in \mathcal{H}_K$ and $h \in span\{\psi_p\}$, minimizing the regularized risk

$$\Lambda((\mathbf{x}_1, y_1, \bar{f}(\mathbf{x}_1)), \cdots, (\mathbf{x}_l, y_l, \bar{f}(\mathbf{x}_l))) + \Omega(\|f\|_K)$$

admits a representation of the form

$$\bar{f}(\mathbf{x}) = \sum_{i=1}^{l} \alpha_i K(\mathbf{x}_i, \mathbf{x}) + \sum_{p=1}^{M} \beta_p \psi_p(\mathbf{x}) \qquad (3.63)$$

with $\alpha_i, \beta_p \in \mathbb{R}$.

In Theorem 3.6, the parametric functions $\{\psi_p\}_{p=1}^{M}$ can be any functions. The simplest parametric function is the constant function $\psi_1(\mathbf{x}) = 1, M = 1$, as in the standard SVM model where the constant function is used to maximize the margin.

In Equation 3.62, the family of parametric functions $\{\psi_p\}_{p=1}^{M}$ does not contribute to the standard regularizer $\|f\|_K^2$. This need not be a major concern if M is sufficiently smaller than l. Here we use $M = 1$ and this parametric function is learned from the whole data, including the labeled and unlabeled data. Therefore, the $l \times M$ matrix $(\psi_p(\mathbf{x}_i))_{ip}$ is a vector whose rank is 1. We denote by $\psi(\mathbf{x})$ this parametric function and by β the corresponding coefficient. Thus, the minimizer of Equation 3.62 is

$$\bar{f}^*(\mathbf{x}) = \sum_{i=1}^{l} \alpha_i^* K(\mathbf{x}_i, \mathbf{x}) + \beta^* \psi(\mathbf{x}) \qquad (3.64)$$

where K is the kernel in the original RKHS $\mathcal{H}_K$.

$\psi(\mathbf{x})$ is obtained by applying the KPCA algorithm [187] to the whole data set. KPCA finds the principal axes in the feature space which carry more variance than any other directions by diagonalizing the covariance matrix $\mathbf{C} = \frac{1}{l+u} \sum_{j=1}^{l+u} \Phi(\mathbf{x}_j)\Phi(\mathbf{x}_j)^\top$, where Φ is a mapping function in the RKHS. To find the principal axes, we solve the eigenvalue problem, $(l+u)\lambda\boldsymbol{\gamma} = K_u\boldsymbol{\gamma}$, where K_u is the kernel used. Let λ denote the largest eigenvalue of K_u and $\boldsymbol{\gamma}$ the corresponding eigenvector. Then the most important principal axis is given by

$$\mathbf{v} = \sum_{i=1}^{l+u} \gamma_i \Phi(\mathbf{x}_i) \qquad (3.65)$$

Usually we normalize $\mathbf{v}$ such that $\|\mathbf{v}\| = 1$. Given the data point $\mathbf{x}$, the projection onto the principal axis is given by $\langle \Phi(\mathbf{x}), \mathbf{v} \rangle$. Let $\psi(\mathbf{x}) = \langle \Phi(\mathbf{x}), \mathbf{v} \rangle =$

$K_u(\mathbf{x}, .)\boldsymbol{\gamma}$. Figure 3.11 shows an illustrative example for the binary classification problem. As shown in this example, $\psi(\mathbf{x})$ might not be the desired classification function. However, $\psi(\mathbf{x})$ is parallel to the desired classification function (the dashed line). They are different up to a constant. Therefore, $\psi(\mathbf{x})$ reflects the geometric structure of the distribution of the data. From this example, it is clear that the data points projected onto the most important principal axis still keep the original neighborhood relationship. In other words, after projection onto the principal axis, similar data points stay close and dissimilar data points are kept far away from each other. In the ideal case of separable binary class problem, we have the following theorem, which says that the similar data points in the feature space are still similar to each other after projected onto the principal axis [96].

THEOREM 3.7
Denote by $\mathcal{C}_i, i = 0, 1$ the set of the data points of each class in the binary class problem. Suppose $\mathcal{C}_i = \{\mathbf{x} | \|\Phi(\mathbf{x}) - \mathbf{c}_i\| \le r_i\}$ and $\|\mathbf{c}_0 - \mathbf{c}_1\| > r_0 + r_1$. For each class, suppose that the data points are uniformly distributed in the sphere of radius r_i. $\|.\|$ denotes the Euclidean norm and $\mathbf{v}$ denotes the principal axis derived from KPCA as defined in Equation 3.65. Then

$$\forall \mathbf{p} \in \mathcal{C}_i, \mathbf{v}^\top \Phi(\mathbf{p}) \in R_i, i = 0, 1$$

where $R_i = [\mu_i - r_i, \mu_i + r_i]$ and $\mu_i = \mathbf{v}^\top \mathbf{c}_i$. Moreover, R_0 and R_1 do not overlap.

Based on the above analysis, the semi-supervised learning is achieved by a 2-step learning process. The first step is to obtain a parametric function $\psi(\mathbf{x})$ from the whole data. Since this parametric function $\psi(\mathbf{x})$ is obtained by KPCA, $\psi(\mathbf{x})$ reflects the geometric structure of the marginal distribution of the data revealed by the whole data. This implements the cluster assumption indirectly. The second step is to solve Equation 3.62 in a new function space to obtain the final classification function.

If $K_u = K$, the final classification function has the form $\bar{f}(\mathbf{x}) = \sum_{i=1}^{l+u} \alpha'_i K(\mathbf{x}_i, \mathbf{x})$ where α'_i is the linear combination of α_i and β. This classification function has the same form as that in [18]; but the methods to obtain it are different. In this case, the parametric function belongs to the original RKHS. Adding $\psi(\mathbf{x})$ does not change the RKHS, but guides the learned classification function toward the desired direction described by $\psi(\mathbf{x})$. If K_u and K are two different kernels, the original RKHS is extended by $\psi(\mathbf{x})$.

The coefficient β^* reflects the weight of the unlabeled data in the learning process. When $\beta^* = 0$, the unlabeled data are not considered at all and this method is fully a supervised learning algorithm. This means that the unlabeled data do not provide any useful information. In other words, the unlabeled data follow the marginal distribution described by the labeled data. When $\beta^* \ne 0$, the unlabeled data provide useful information about the

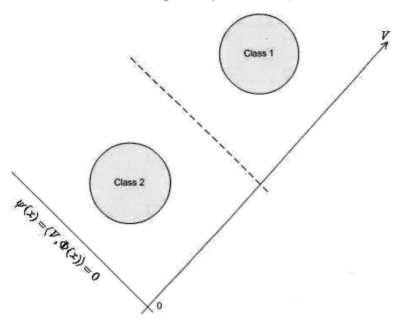

FIGURE 3.11: Illustration of KPCA in the two dimensions.

marginal distribution of the data and the geometric structure of the marginal distribution revealed by the unlabeled data is incorporated into the learning.

To learn the final classification function, we substitute Equation 3.64 into Equation 3.62 to obtain an objective function of α_i^* and β^*. The solution of α_i^* and β^* depends on the loss function. Different loss functions L result in different algorithms. We now discuss two typical loss functions: the squared loss for RLS and the hinge loss for SVM. For the squared loss function, we obtain the explicit form of α_i^* and β^*. In the following analysis of this section, we use K interchangeably to denote the kernel function or the kernel matrix.

3.11.3 Semiparametric Regularized Least Squares

We first outline the RLS approach which applies to the binary classification and the regression problem. The classic RLS algorithm is a supervised method where we solve:

$$f^* = \arg\min_{f \in \mathcal{H}_K} C \sum_{i=1}^{l} (y_i - f(\mathbf{x}_i))^2 + \gamma \|f\|_K^2$$

where C and γ are constants.

According to Theorem 3.5, the solution is of the following form:

$$f^*(\mathbf{x}) = \sum_{i=1}^{l} \alpha_i^* K(\mathbf{x}_i, \mathbf{x})$$

Substituting this solution in the problem above, we arrive at the following differentiable objective function of the l-dimensional variable $\boldsymbol{\alpha} = [\alpha_1 \cdots \alpha_l]^\top$:

$$\boldsymbol{\alpha}^* = \arg\min C(Y - K\boldsymbol{\alpha})^\top (Y - K\boldsymbol{\alpha}) + \gamma \boldsymbol{\alpha}^\top K \boldsymbol{\alpha}$$

where K is the $l \times l$ kernel matrix $K_{ij} = K(\mathbf{x}_i, \mathbf{x}_j)$ and $\mathbf{Y}$ is the label vector $\mathbf{Y} = [y_1 \cdots y_l]^\top$.

The derivative of the objective function over $\boldsymbol{\alpha}$ vanishes at the minimizer

$$C(KK\boldsymbol{\alpha}^* - K\mathbf{Y}) + \gamma K\boldsymbol{\alpha}^* = 0$$

which leads to the following solution:

$$\boldsymbol{\alpha}^* = (CK + \gamma\mathbf{I})^{-1}C\mathbf{Y}$$

The semiparametric RLS algorithm solves the optimization problem in Equation 3.62 with the squared loss function:

$$\bar{f}^* = \arg\min_{\bar{f}} C \sum_{i=1}^{l} (y_i - \bar{f}(\mathbf{x}_i))^2 + \gamma \|f\|_K^2 \qquad (3.66)$$

where $\bar{f} := f + h$ with $f \in \mathcal{H}_K$ and $h \in span\{\psi\}$.

According to Theorem 3.6, the solution has the form

$$\bar{f}^* = \sum_{i=1}^{l} \alpha_i^* K(\mathbf{x}_i, \mathbf{x}) + \beta^* \psi(\mathbf{x})$$

Substituting this form in Equation 3.66, we arrive at the following objective function of the l-dimensional variable $\boldsymbol{\alpha} = [\alpha_1 \cdots \alpha_l]^\top$ and β:

$$(\boldsymbol{\alpha}^*, \beta^*) = \arg\min C\boldsymbol{\delta}^\top \boldsymbol{\delta} + \gamma \boldsymbol{\alpha}^\top K \boldsymbol{\alpha}$$

where $\boldsymbol{\delta} = Y - K\boldsymbol{\alpha} - \beta\boldsymbol{\psi}$, K is the $l \times l$ kernel matrix $K_{ij} = K(\mathbf{x}_i, \mathbf{x}_j)$, $\mathbf{Y}$ is the label vector $\mathbf{Y} = [y_1 \cdots y_l]^\top$, and $\boldsymbol{\psi}$ is the vector $\boldsymbol{\psi} = [\psi(\mathbf{x}_1) \cdots \psi(\mathbf{x}_l)]^\top$. The derivatives of the objective function over $\boldsymbol{\alpha}$ and β vanish at the minimizer:

$$C(KK\boldsymbol{\alpha}^* + \beta^* K\boldsymbol{\psi} - K\mathbf{Y}) + \gamma K\boldsymbol{\alpha}^* = 0$$
$$\boldsymbol{\psi}^\top K\boldsymbol{\alpha}^* + \beta^*\boldsymbol{\psi}^\top\boldsymbol{\psi} - \boldsymbol{\psi}^\top Y = 0$$

which lead to the following solution:

$$\boldsymbol{\alpha}^* = C(\gamma\mathbf{I} - \frac{C\boldsymbol{\psi}\boldsymbol{\psi}^\top K}{\boldsymbol{\psi}^\top\boldsymbol{\psi}} + CK)^{-1}(\mathbf{I} - \frac{\boldsymbol{\psi}\boldsymbol{\psi}^\top}{\boldsymbol{\psi}^\top\boldsymbol{\psi}})\mathbf{Y} \qquad (3.67)$$

$$\beta^* = \frac{\boldsymbol{\psi}^\top Y - \boldsymbol{\psi}^\top K\boldsymbol{\alpha}^*}{\boldsymbol{\psi}^\top\boldsymbol{\psi}}$$

3.11.4 Semiparametric Regularized Support Vector Machines

We outline the SVM approach to the binary classification problem.

In the binary classification problem, the classic SVM attempts to solve the following optimization problem on the labeled data.

$$\min \ \frac{1}{2}\|\mathbf{w}\|^2 + C\sum_{i=1}^{l}\xi_i \tag{3.68}$$

$$s.t. \ \ y_i\{\langle \mathbf{w}, \Phi(\mathbf{x}_i)\rangle + b\} \geq 1 - \xi_i$$

$$\xi_i \geq 0 \quad i = 1,\cdots,l$$

where Φ is a nonlinear mapping function determined by the kernel and b is a regularized term.

Again, the solution is given by

$$f^*(\mathbf{x}) = \langle \mathbf{w}^*, \Phi(\mathbf{x})\rangle + b^* = \sum_{i=1}^{l}\alpha_i^* K(\mathbf{x}_i, \mathbf{x}) + b^*$$

To solve Equation 3.68, we introduce one Lagrange multiplier for each constraint in Equation 3.68 using the Lagrange multipliers technique and obtain a quadratic dual problem of the Lagrange multipliers.

$$\min \ \frac{1}{2}\sum_{i,j=1}^{l} y_i y_j \mu_i \mu_j K(\mathbf{x}_i, \mathbf{x}_j) - \sum_{i=1}^{l}\mu_i \tag{3.69}$$

$$s.t. \ \sum_{i=1}^{l}\mu_i y_i = 0$$

$$0 \leq \mu_i \leq C \quad i = 1,\cdots,l$$

where μ_i is the Lagrange multiplier associated with the i-th constraint in Equation 3.68.

We have $\mathbf{w}^* = \sum_{i=1}^{l}\mu_i y_i \Phi(\mathbf{x}_i)$ from the solution to Equation 3.69. Note that the following conditions must be satisfied according to the Kuhn-Tucker theorem [209]:

$$\mu_i(y_i(\langle \mathbf{w}, \Phi(\mathbf{x}_i)\rangle + b) + \xi_i - 1) = 0 \quad i = 1,\cdots,l \tag{3.70}$$

The optimal solution of b is determined by the above conditions.

Therefore, the solution is given by

$$f^*(\mathbf{x}) = \sum_{i=1}^{l}\alpha_i^* K(\mathbf{x}_i, \mathbf{x}) + b^*$$

where $\alpha_i^* = \mu_i y_i$.

The semiparametric SVM algorithm solves the optimization problem in Equation 3.62 with the hinge loss function:

$$\min \ \frac{1}{2}\|\mathbf{w}\|^2 + C\sum_{i=1}^{l}\xi_i \qquad (3.71)$$

$$s.t. \ \ y_i\{\langle \mathbf{w}, \Phi(\mathbf{x}_i)\rangle + b + \beta\psi(\mathbf{x}_i)\} \geq 1 - \xi_i$$

$$\xi_i \geq 0 \quad i = 1, \cdots, l$$

As in the classic SVM, we consider the Lagrange dual problem for Equation 3.71:

$$\min \ \frac{1}{2}\sum_{i,j=1}^{l} y_i y_j \mu_i \mu_j K(\mathbf{x}_i, \mathbf{x}_j) - \sum_{i=1}^{l}\mu_i \qquad (3.72)$$

$$s.t. \ \ \sum_{i=1}^{l}\mu_i y_i = 0$$

$$\sum_{i=1}^{l}\mu_i y_i \psi(\mathbf{x}_i) = 0$$

$$0 \leq \mu_i \leq C \quad i = 1, \cdots, l$$

where μ_i is the Lagrange multiplier associated with the i-th constraint in Equation 3.71. The semiparametric SVM dual problem Equation 3.72 is the same as the SVM dual problem Equation 3.69 except for one more constraint introduced by the parametric function $\psi(\mathbf{x})$. As in the classic SVM, the following conditions must be satisfied:

$$\mu_i(y_i(\langle \mathbf{w}, \Phi(\mathbf{x}_i)\rangle + b + \beta\psi(\mathbf{x}_i)) + \xi_i - 1) = 0 \qquad (3.73)$$

We have $\mathbf{w}^* = \sum_{i=1}^{l}\mu_i y_i \Phi(\mathbf{x}_i)$ from the solution to Equation 3.72. This is the same as that in the SVM.

The optimal solution of b^* and β^* is determined by Equation 3.73. If the number of the Lagrange multipliers satisfying $0 < \mu_i < C$ is no less than two, we may determine b^* and β^* by solving two linear equations corresponding to any two of them in Equation 3.73 since the corresponding slack variable ξ_i is zero. In the case that the number of the Lagrange multipliers satisfying $0 < \mu_i < C$ is less than two, b^* and β^* are determined by solving the following optimization problem derived from Equation 3.73:

$$\min \ b^2 + \beta^2 \qquad (3.74)$$

$$s.t. \ \ y_i\{\langle \mathbf{w}, \Phi(\mathbf{x}_i)\rangle + b + \beta\psi(\mathbf{x}_i)\} \geq 1$$

$$\text{if } \mu_i = 0$$

$$y_i\{\langle \mathbf{w}, \Phi(\mathbf{x}_i)\rangle + b + \beta\psi(\mathbf{x}_i)\} = 1$$

$$\text{if } 0 < \mu_i < C$$

The final decision function is

$$\bar{f}^*(\mathbf{x}) = \sum_{i=1}^{l} \alpha_i^* K(\mathbf{x}_i, \mathbf{x}) + \beta^* \psi(\mathbf{x}) + b^*$$

where $\alpha_i^* = \mu_i y_i$. The semiparametric SVM can be implemented by using a standard quadratic programming problem solver.

3.11.5 Semiparametric Regularization Algorithm

Based on the above analysis, the semiparametric regularization algorithm is summarized in Algorithm 4.

Algorithm 4 Semiparametric Regularization Algorithm

Input: l labeled data points $(\mathbf{x}_i, y_i), 1 \leq i \leq l, y_i = \pm 1$ and u unlabeled data points $\mathbf{x}_i, l + 1 \leq i \leq l + u$

Output: Estimated function $\bar{f}^*(\mathbf{x}) = \sum_{i=1}^{l} \alpha_i^* K(\mathbf{x}_i, \mathbf{x}) + \beta^* \psi(\mathbf{x})$ for SpRLS or $\bar{f}^*(\mathbf{x}) = \sum_{i=1}^{l} \alpha_i^* K(\mathbf{x}_i, \mathbf{x}) + \beta^* \psi(\mathbf{x}) + b^*$ for SpSVM

Method:

1: Choose the kernel K_u and apply KPCA to the whole data to obtain the parametric function $\psi(\mathbf{x}) = \sum_{i=1}^{l+u} \gamma_i K_u(\mathbf{x}_i, \mathbf{x})$.

2: Choose the kernel K and solve Equation 3.67 for SpRLS or Equations 3.72 and 3.74 for SpSVM.

3.11.6 Transductive Learning and Semi-Supervised Learning

Transductive learning refers to the learning scenario where we are given a set of labeled training data and another set of unlabeled training data, and the goal is to predict the labels for the unlabeled training data without necessarily inducing the classifier function. Semi-supervised learning refers to the scenario where we are given a set of labeled training data and another set of unlabeled training data, and the goal is to induce the classifier function using both the labeled and unlabeled training data such that for any unseen data we can use the induced function to predict the label of the unseen data.

Transductive learning only works on the labeled and unlabeled training data and cannot handle unseen data. Out-of-sample extension is already a serious limitation for transductive learning. In contrast to transductive learning, inductive learning can handle unseen data. Semi-supervised learning can be either transductive or inductive. Many existing graph-based semi-supervised learning methods are transductive in nature since the classification function is only defined on the labeled and unlabeled training data. One reason is that they perform the semi-supervised learning only on the graph where the nodes are the labeled and unlabeled data in the training set, not on the whole space.

In this presented approach, the decision function Equation 3.64 is defined over the whole $\mathcal{X}$ space. Therefore, this approach is inductive in nature and can extend to the out-of-sample data.

3.11.7   Comparisons with Other Methods

In the literature, many existing semi-supervised learning methods rely on the *cluster assumption* directly or indirectly and exploit the regularization principle by considering additional regularization terms on the unlabeled data. Belkin et al [18] propose a manifold regularization approach where the geometric structure of the marginal distribution is extracted using the graph Laplacian associated with the data. They consider the following regularization term:

$$\sum_{i,j=1}^{l+u} (f(\mathbf{x}_i) - f(\mathbf{x}_j))^2 W_{ij} = \mathbf{f}^\top \mathbf{L} \mathbf{f} \qquad (3.75)$$

where W_{ij} are edge weights in the data adjacency graph and $\mathbf{L}$ is the graph Laplacian given by $\mathbf{L} = \mathbf{D} - \mathbf{W}$. Here, the diagonal matrix $\mathbf{D}$ is given by $D_{ii} = \sum_{j=1}^{l+u} W_{ij}$. The incorporation of this regularization term leads to the following optimization problem:

$$f^* = \arg \min_{f \in \mathcal{H}_K} C \sum_{i=1}^{l} L(\mathbf{x}_i, y_i, f(\mathbf{x}_i)) + \gamma \|f\|_K^2 + \mathbf{f}^\top \mathbf{L} \mathbf{f}$$

Equation 3.75 attempts to give the nearby points (large W_{ij}) in the graph similar labels. However, the issue is that Equation 3.75 tends to give similar labels for points i and j as long as $W_{ij} > 0$. In other words, dissimilar points might have similar labels. Therefore, their approach depends on the neighborhood graph constructed from the data. Similarly, Zhu et al [259] minimize Equation 3.75 as an energy function.

The semiparametric regularization based semi-supervised learning approach presented in this section exploits the cluster assumption by the parametric function $\psi(\mathbf{x})$. Learned from the whole data, this parametric function reflects the geometric structure of the marginal distribution of the data. Different from the manifold regularization approach, this approach uses a parametric function obtained from the whole data to describe the geometric structure of the marginal distribution. Similar to the manifold regularization approach, this approach obtains the same form of the classification function if we use the same kernel ($K = K_u$) in the 2-step learning process. However, the methods to obtain the expansion coefficients are different.

Sindhwani et al [189] derive a modified kernel defined in the same space of the functions as the original RKHS, but with a different norm. Here we warp an RKHS in a different way. We extend the original RKHS by including the parametric function without changing the norm such that the learned decision function reflects the data distribution. In some cases, this parametric function belongs to the original RKHS and thus the RKHS is unchanged. However, the learned classification function still reflects the data distribution since the classification function has a preference to the parametric function according to Equation 3.64.

The parametric function $\psi(\mathbf{x})$ learned by KPCA can be incorporated into the supervised learning to separate different classes very well for the binary classification problem. For the multiclass problem, KPCA cannot separate different classes very well because some classes overlap after projection onto the principal axis. That is why we focus on the binary class problem in this approach. The evaluations of this approach as well as the superiority to the peers methods from the state-of-the-art machine learning literature are reported and demonstrated in [96].

3.12 Summary

In this chapter we have introduced the commonly used and recently developed statistical learning and mining theory and techniques in the context of multimedia data mining. We have studied and introduced the two well-known paradigms of statistical learning in the recent multimedia data mining literature — the generative learning models and the discriminative learning models. In the former we focus on the probabilistic inference based learning methods, including Bayesian networks, probabilistic latent semantic analysis, latent Dirichlet allocation, and hierarchical Dirichlet process for discrete data analysis, and we have briefly reviewed their applications in multimedia data mining. In the latter we focus on the support vector machines as well as the recently developed maximum margin learning with structured output space and the boosting theory to combine a series of weak learners to build up a strong learner. We then have studied and introduced two recently developed statistical learning paradigms that have been extensively applied in the recent multimedia data mining literature — multiple instance learning and semi-supervised learning. These statistical learning methods are the theoretical foundation of all multimedia data mining tasks. They are extensively applied in the subsequent chapters in Part III of this book.

Chapter 4

Soft Computing Based Theory and Techniques

4.1 Introduction

In many multimedia data mining applications, it is often required to make a decision in an imprecise and uncertain environment. For example, in the application of mining an image database with a query image of green trees, given an image in the database that is about a pond with a bank of earth and a few green bushes, is this image considered as a match to the query? Certainly this image is not a perfect match to the query, but, on the other hand, it is also not an absolute mismatch to the query. Problems like this example, as well as many others, have intrinsic imprecision and uncertainty that cannot be neglected in decision making. Traditional intelligent systems fail to solve such problems, as they attempt to use *Hard Computing* techniques. In contrast, a *Soft Computing* methodology implies cooperative activities rather than autonomous ones, resulting in new computing paradigms such as fuzzy logic, neural networks, and evolutionary computation. Consequently, soft computing opens up a new research direction for problem solving that is difficult to achieve using traditional hard computing approaches.

Technically, soft computing includes specific research areas such as fuzzy logic, neural networks, genetic algorithms, and chaos theory. Intrinsically, soft computing is developed to deal with pervasive imprecision and uncertainty of real-world problems. Unlike traditional hard computing, soft computing is capable of tolerating imprecision, uncertainty, and partial truth without loss of performance and effectiveness for the end user. The guiding principle of soft computing is to exploit the tolerance for imprecision, uncertainty, and partial truth to achieve a required tractability, robustness, and low solution cost. We can easily come to the conclusion that precision has a cost. Therefore, in order to solve a problem with an acceptable cost, we need to aim at a decision with only the necessary degree of precision, not exceeding the requirements.

In soft computing, fuzzy logic is the kernel. The principal advantage of fuzzy logic is the robustness to its interpolative reasoning mechanism. Within soft computing, fuzzy logic is mainly concerned with imprecision and approximate reasoning, neural networks with learning, genetic algorithms with global optimization and search, and chaos theory with nonlinear dynamics. Each of these computational paradigms provides us with complementary reasoning and searching methods to solve complex, real-world problems. The interrelations between these paradigms of soft computing contribute to the theoretical foundation of *Hybrid Intelligent Systems*. The use of hybrid intelligent systems leads to the development of numerous manufacturing systems, multimedia systems, intelligent robots, and trading systems, well beyond the scope of multimedia data mining.

4.2 Characteristics of the Paradigms of Soft Computing

Different paradigms of soft computing can be used independently and more often in combination. In soft computing, fuzzy logic plays a unique role. Fuzzy sets are used as a universal approximator, which is often paramount for modeling unknown objects. However, fuzzy logic in its pure form may not necessarily always be useful for easily constructing an intelligent system. For example, when a designer does not have sufficient prior information (knowledge) about the system, the development of acceptable fuzzy rules becomes impossible; further, as the complexity of the system increases, it becomes difficult to specify a correct set of rules and membership functions for adequately and correctly describing the behavior of the system. Fuzzy systems also have the disadvantage of the inability to automatically extract additional knowledge from the experience and to automatically correct and improve the fuzzy rules of the system.

Another important paradigm of soft computing is neural networks. Artificial neural networks, as a parallel, fine-grained implementation of non-linear static or dynamic systems, were originally developed as a parallel computational model. A very important advantage of these networks is their adaptive capability, where "learning by example" replaces the traditional "programming" in problem solving. Another important advantage is the intrinsic parallelism that allows fast computations. Artificial neural networks are a viable computational model for a wide variety of problems, including pattern classification, speech synthesis and recognition, curve fitting, approximation, image compression, associative memory, and modeling and control of non-linear unknown systems, in addition to the application of multimedia data mining. The third advantage of artificial neural networks is the generalization capability, which allows correct classification of new patterns. A significant disadvantage of artificial neural networks is their poor interpretability. One of the main criticisms addressed to neural networks concerns their black box nature.

Evolutionary computing is a revolutionary paradigm for optimization. One component of evolutionary computing — genetic algorithms — studies the al-

Table 4.1: Comparative characteristics of the components of soft computing. Reprint from [8] ©2001 World Scientific.

	Fuzzy sets	**Artificial neural networks**	**Evolutionary computing, Genetic algorithms**
Weakness	Knowledge acquisition; Learning	Black box interpretability	Coding; Computational speed
Strengths	Interpretability; Transparency; Plausibility; Modeling; Reasoning; Tolerance to imprecision	Learning; Adaptation; Fault tolerance; Curve fitting; Generalization ability; Approximation ability	Computational efficiency; Global optimization

gorithms for global optimization. Genetic algorithms are based on the mechanisms of natural selection and genetics. One advantage of genetic algorithms is that they effectively implement a parallel, multi-criteria search. The mechanism of genetic algorithms is simple. Simplicity of operations and powerful computational effect are the two main principles for designing effective genetic algorithms. The disadvantages include the convergence issue and the lack of strong theoretic foundation. The requirement of coding the domain variables into bit strings also seems to be a drawback of genetic algorithms. In addition, the computational speed of genetic algorithms is typically low.

Table 4.1 summarizes the comparative characteristics of the different paradigms of soft computing. For each paradigm of soft computing, there are appropriate problems where this paradigm is typically applied.

4.3 Fuzzy Set Theory

In this section, we give an introduction to fuzzy set theory, fuzzy logic, and their applications in multimedia data mining.

4.3.1 Basic Concepts and Properties of Fuzzy Sets

DEFINITION 4.1 *Let X be a classic set of objects, called the* universe, *with the generic elements denoted as x. The membership of a classic subset*

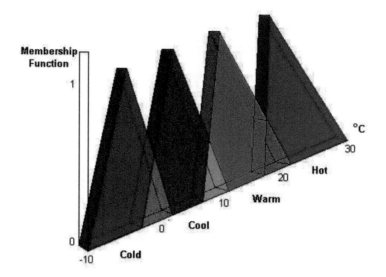

FIGURE 4.1: Fuzzy set to characterize the temperature of a room.

A of X is often considered as a characteristic function μ_A mapped from X to {0,1} such that

$$\mu_A(x) = \begin{cases} 1 \text{ iff } x \in A \\ 0 \text{ iff } x \notin A \end{cases}$$

where {0,1} is called a valuation set; 1 indicates membership while 0 indicates non-membership.

If the valuation set is allowed to be in the real interval $[0,1]$, A is called a fuzzy set. $\mu_A(x)$ is the grade of membership of x in A:

$$\mu_A : X \longrightarrow [0,1]$$

The closer the value of $\mu_A(x)$ is to 1, the more x belongs to A. A is completely characterized by the set of the pair:

$$A = \{(x, \mu_A(x)), x \in X\}$$

Solutions to many real-world problems can be developed more accurately using fuzzy set theory. Figure 4.1 shows an example regarding how fuzzy set representation is used to describe the natural drift of temperature.

DEFINITION 4.2   *Two fuzzy sets A and B are said to be equal, $A = B$, if and only if $\forall x \in X \ \mu_A(x) = \mu_B(x)$.*

In the case where universe $\mathbf{X}$ is infinite, it is desirable to represent fuzzy sets in an analytical form, which describes the mathematical membership functions. There are several mathematical functions that are frequently used as the membership functions in fuzzy set theory and practice. For example, a Gaussian-like function is typically used for the representation of the membership function as follows:

$$\mu_A(x) = c\exp(-\frac{(x-a)^2}{b})$$

which is defined by three parameters, a, b, and c. Figure 4.2 summarizes the graphical and analytical representations of frequently used membership functions.

An appropriate construction of the membership function for a specific fuzzy set is the problem of *knowledge engineering* [125]. There are many methods for an appropriate estimation of a membership function. They can be categorized as follows:

1. Membership functions based on heuristics

2. Membership functions based on reliability concepts with respect to the specific problem

3. Membership functions based on a certain theoretic foundation

4. Neural networks based construction of membership functions

The following rules which are common and valid in the classic set theory also apply to fuzzy set theory.

- De Morgan's law:
$$\overline{A \cap B} = \overline{A} \cup \overline{B}$$
and
$$\overline{A \cup B} = \overline{A} \cap \overline{B}$$

- Associativity:
$$(A \cup B) \cup C = A \cup (B \cup C)$$
and
$$(A \cap B) \cap C = A \cap (B \cap C)$$

- Commutativity:
$$A \cup B = B \cup A$$
and
$$A \cap B = B \cap A$$

- Distributivity:
$$A \cup (B \cap C) = (A \cup B) \cap (A \cup C)$$
and
$$A \cap (B \cup C) = (A \cap B) \cup (A \cap C)$$

Multimedia Data Mining

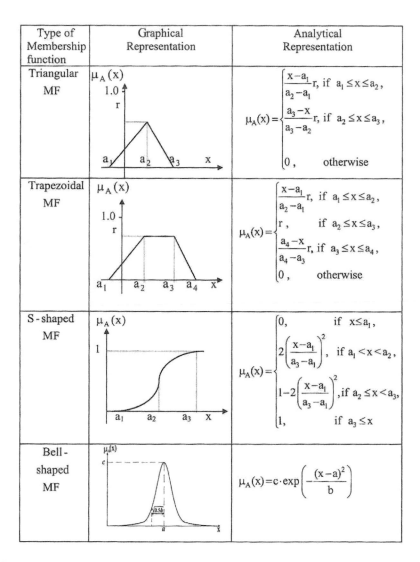

Type of Membership function	Graphical Representation	Analytical Representation
Triangular MF	$\mu_A(x)$	$\mu_A(x)=\begin{cases}\dfrac{x-a_1}{a_2-a_1}r, & \text{if } a_1\le x\le a_2,\\[6pt]\dfrac{a_3-x}{a_3-a_2}r, & \text{if } a_2\le x\le a_3,\\[6pt]0, & \text{otherwise}\end{cases}$
Trapezoidal MF	$\mu_A(x)$	$\mu_A(x)=\begin{cases}\dfrac{x-a_1}{a_2-a_1}r, & \text{if } a_1\le x\le a_2,\\[6pt]r, & \text{if } a_2\le x\le a_3,\\[6pt]\dfrac{a_4-x}{a_4-a_3}r, & \text{if } a_3\le x\le a_4,\\[6pt]0, & \text{otherwise}\end{cases}$
S-shaped MF	$\mu_A(x)$	$\mu_A(x)=\begin{cases}0, & \text{if } x\le a_1,\\[6pt]2\left(\dfrac{x-a_1}{a_3-a_1}\right)^2, & \text{if } a_1<x<a_2,\\[6pt]1-2\left(\dfrac{x-a_1}{a_3-a_1}\right)^2, & \text{if } a_2\le x<a_3,\\[6pt]1, & \text{if } a_3\le x\end{cases}$
Bell-shaped MF	$\mu_A(x)$	$\mu_A(x)=c\cdot\exp\left(-\dfrac{(x-a)^2}{b}\right)$

FIGURE 4.2: Typical membership functions. Reprint from [8] ©2001 World Scientific.

4.3.2 Fuzzy Logic and Fuzzy Inference Rules

In this section fuzzy logic is reviewed in a narrow sense as a direct extension and generalization of multi-valued logic. According to one of the most widely accepted definitions, logic is an analysis of methods of reasoning; in studying these methods, logic is mainly taken in the form, not in the content, of the arguments used in a reasoning process. Here the main issue is to establish whether the truth of the consequence can be inferred from the truth of premises. Systematic formulation of the correct approaches to reasoning is one of the main issues in logic.

Let us define the semantic truth function of fuzzy logic. Let P be a statement and T(P) be its truth value, where $T(P) \in [0,1]$. Negation values of the statement P are defined as $T(\neg P) = 1 - T(P)$. The implication connective is always defined as

$$T(P \rightarrow Q) = T(\neg P \vee Q)$$

and the equivalence is always defined as

$$T(P \leftrightarrow Q) = T[(P \rightarrow Q) \wedge (Q \rightarrow P)]$$

Based on the above definitions, we further define the basic connectives of fuzzy logic as follows.

- $T(P \vee Q) = \max(T(P), T(Q))$

- $T(P \wedge Q) = \min(T(P), T(Q))$

- $T(P \vee (P \wedge Q)) = T(P)$

- $T(P \wedge (P \vee Q)) = T(P)$

- $T(\neg(P \wedge Q)) = T(\neg P \vee \neg Q)$

- $T(\neg(P \vee Q)) = T(\neg P \wedge \neg Q)$

It is shown that multi-valued logic is the fuzzification of the traditional propositional calculus (in the sense of the extension principle). Here each proposition P is assigned a normalized fuzzy set in [0,1]; i.e., the pair $\{\mu_P(0), \mu_P(1)\}$ is interpreted as the degree of false or true, respectively. Since the logical connectives of the standard propositional calculus are functionals of truth, i.e., they are represented as functions, they can be fuzzified.

Let A and B be fuzzy sets of the subsets of the non-fuzzy universe U; in fuzzy set theory it is known that A is a subset of B iff $\mu_A \leq \mu_B$, i.e., $\forall x \in U$, $\mu_A(x) \leq \mu_B(x)$.

In fuzzy set theory, great attention is paid to the development of fuzzy conditional inference rules. This is connected to the natural language understanding where it is necessary to have a certain number of fuzzy concepts; therefore, we must ensure that the inference of the logic is made such that the preconditions and the conclusions both may contain such fuzzy concepts. It is shown that there is a huge variety of ways to formulate the rules for such inferences. However, such inferences cannot be satisfactorily formulated using the classic Boolean logic. In other words, here we need to use multi-valued logical systems. The conceptual principle in the formulation of the fuzzy rules is the Modus Ponens inference rule that states: IF$(\alpha \to \beta)$ is true and α is true, THEN β must also be true.

The methodological foundation for this formulation is the compositional rule suggested by Zadeh [231, 232]. Using this rule, he has formulated the inference rules in which both the logical preconditions and consequences are conditional propositions, including the fuzzy concepts.

4.3.3 Fuzzy Set Application in Multimedia Data Mining

In multimedia data mining, fuzzy set theory can be used to address the typical uncertainty and imperfection in the representation and processing of multimedia data, such as image segmentation, feature representation, and feature matching. Here we give one such application in image feature representation as an example in multimedia data mining.

In image data mining, the image feature representation is the very first step for any knowledge discovery in an image database. In this example, we show how different image features may be represented appropriately using the fuzzy set theory.

In Section 2.4.5.2, we have shown how to use fuzzy logic to represent the color features. Here we show the fuzzy representation of texture and shape features for a region in an image. Similar to the color feature, the fuzzification of the texture and shape features also brings a crucial improvement into the region representation of an image, as the fuzzy features naturally characterize the gradual transition between regions within an image. In the following proposed representation scheme, a fuzzy feature set assigns weights, called the degree of membership, to feature vectors of each image block in the feature space. As a result, the feature vector of a block belongs to multiple regions with different degrees of membership as opposed to the classic region representation, in which a feature vector belongs to exactly one region. We first discuss the fuzzy representation of the texture feature, and then discuss that of the shape feature.

We take each region as a fuzzy set of blocks. In order to propose a unified approach consistent with the fuzzy color histogram representation described in Section 2.4.5.2, we again use the Cauchy function to be the fuzzy membership function, i.e.,

$$\mu_i(f) = \frac{1}{1 + (\frac{d(f,\hat{f}_i)}{\sigma})^\alpha} \tag{4.1}$$

where $f \in R^k$ is the texture feature vector of each block with k as the dimensionality of the feature vector; $\hat{f}_i$ is the average texture feature vector of region i; d is the Euclidean distance between $\hat{f}_i$ and any feature f; and σ

represents the average distance for texture features among the cluster centers obtained from the k-means algorithm. σ is defined as:

$$\sigma = \frac{2}{C(C-1)} \sum_{i=1}^{C-1} \sum_{k=i+1}^{C} \|\hat{f}_i - \hat{f}_k\| \tag{4.2}$$

where C is the number of regions in a segmented image, and $\hat{f}_i$ is the average texture feature vector of region i.

With this block membership function, the fuzzified texture property of region i is represented as

$$\vec{f}_i^T = \sum_{f \in U^T} f \mu_i(f) \tag{4.3}$$

where U^T is the feature space composed of texture features of all blocks.

Based on the fuzzy membership function $\mu_i(f)$ obtained in a similar fashion, we also fuzzify the p-th order inertia as the shape property representation of region i as:

$$l(i,p) = \frac{\sum_{f \in U^S} [(f_x - \hat{x})^2 + (f_y - \hat{y})^2]^{p/2} \mu_i(f)}{[N]^{1+p/2}} \tag{4.4}$$

where f_x and f_y are the x and y coordinates of the block with the shape feature f, respectively; $\hat{x}$ and $\hat{y}$ are the x and y central coordinates of region i, respectively; and N is the number of blocks in an image and U^S is the block feature space of the images. Based on Equation 4.4, we have obtained the fuzzy representation for the shape feature of each region, denoted as $\vec{f}_i^S$.

4.4 Artificial Neural Networks

Historically, in order to "simulate" the biological systems to make non-symbolic computations, different mathematical models were suggested. The artificial neural network is one such model that has shown great promise and thus attracted much attention in the literature.

4.4.1 Basic Architectures of Neural Networks

Neurons represent a special type of nervous cells in the organism, having electric activities. These cells are mainly intended for the operative control of the organism. A neuron consists of cell bodies, which are enveloped in the membrane. A neuron also has dendrites and axons, which are its inputs and outputs. Axons of neurons join dendrites of other neurons through synaptic contacts. Input signals of the dendrite tree are weighted and added in the

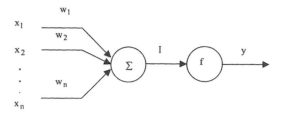

FIGURE 4.3: Mathematical model of a neuron. Reprint from [8] ©2001 World Scientific.

cell body and formed in the axon, where the output signal is generated. The signal's intensity, consequently, is a function of a weighted sum of the input signal. The output signal is passed through the branches of the axon and reaches the synapses. Through the synapses the signal is transformed into a new input signal of the neighboring neurons. This input signal can be either positive or negative, depending upon the type of the synapses.

The mathematical model of the neuron that is usually utilized under the simulation of the neural network is represented in Figure 4.3. The neuron receives a set of input signals $x_1, x_2, ..., x_n$ (i.e., vector X) which usually are output signals of other neurons. Each input signal is multiplied by a corresponding connection weight w — analogue of the synapse's efficiency. Weighted input signals come to the summation module corresponding to the cell body, where their algebraic summation is executed and the excitement level of the neuron is determined:

$$I = \sum_{1=1}^{n} x_i W_i$$

The output signal of a neuron is determined by conducting the excitement level through the function f, called the activation function:

$$y = f(I - \theta)$$

where θ is the threshold of the neuron. Usually the following activation functions are used as function f:

- Linear function (see Figure 4.4),

$$y = kI, \quad k = const$$

- binary (threshold) function (see Figure 4.5),

$$y = \begin{cases} 1 \text{ if } I \geq \theta \\ 0 \text{ if } I < \theta \end{cases}$$

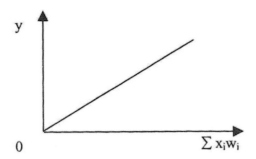

FIGURE 4.4: Linear function. Reprint from [8] ©2001 World Scientific.

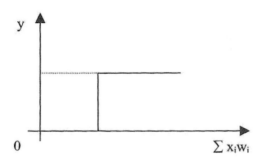

FIGURE 4.5: Binary function. Reprint from [8] ©2001 World Scientific.

- sigmoid function (see Figure 4.6),

$$y = \frac{1}{1 + \exp^{-I}}$$

The totality of the neurons, connected with each other and with the environment, forms the neural network. The input vector comes to the network by activating the input neurons. A set of input signals $x_1, x_2, ..., x_n$ of a network's neurons is called the vector of the input activeness. Connection weights of neurons are represented in the form of a matrix W, the element w_{ij} of which is the connection weight between the i-th and the j-th neurons. During the network functioning process, the input vector is transformed into output one; i.e., a certain information processing is performed. The computational power

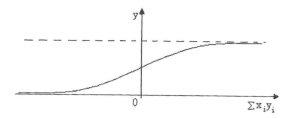

FIGURE 4.6: Sigmoid function. Reprint from [8] ©2001 World Scientific.

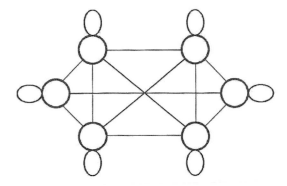

FIGURE 4.7: A fully connected neural network. Reprint from [8] ©2001 World Scientific.

of the network consequently solves problems with its connections. Connections link the inputs of a neuron with the outputs of others. The connection strengths are given by the weight coefficients.

The network's architecture is represented by the order of the connections. Two frequently used network types are the fully-connected networks and the hierarchical networks. In a fully connected architecture, all of its elements are connected with each other. The output of every neuron is connected with the inputs of all others and its own input. The number of the connections in a fully-connected neural network is equal to $v \times v$, with v links for each neuron (see Figure 4.7).

In the hierarchical architecture, a neural network may be differentiated by the neurons grouped into particular layers or levels. Each neuron in any hidden layer is connected with every neuron in the previous and the next layers. There are two special layers in the hierarchical networks. Those layers have contacts and interact with the environment (see Figure 4.8).

In terms of the signal transference direction in the networks, they are categorized into the networks without feedback loops (called feed-forward networks) and the networks with feedback loops (called either feedback or recurrent

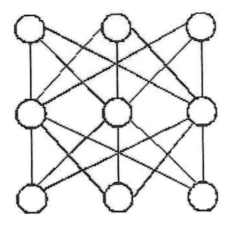

FIGURE 4.8: A hierarchical neural network. Reprint from [8] ©2001 World Scientific.

networks).

In feed-forward networks the neurons of each layer receive signals either from the environment or from neurons of the previous layer and pass their outputs either to the environment or to neurons of the next layer (see Figure 4.9). In recurrent networks (Figure 4.10) neurons of a particular layer may also receive signals from themselves and from other neurons of the layer. Thus, unlike non-recurrent networks, the values of the output signals in a recurrent neural network may be determined only if (besides the current value of the input signals and the weights of the corresponding connections) there is information available about values of the outputs of the neurons in the previous step of the time. This means that such a network possesses elements of memory that allow it to keep information about the outputs' state from some time interval. That is why recurrent networks can model the associative memory. The associative memory is content-addressable. When an incomplete or a corrupted vector comes to such a network, it can retrieve the correct vector.

A non-recurrent (feed-forward) network has no feedback connections. In this network topology neurons of the i-th layer receive signals from the environment (when $i = 1$) or from the neurons of the previous layer, i.e., the $(i - 1)$-th layer (when $i > 1$), and pass their outputs to the neurons of the next $(i + 1)$-th layer or to the environment (when i is the last layer).

The hierarchical non-recurrent network may be single-layer or multi-layer. A non-recurrent network containing one input and one output layer, respectively, usually is called a single-layer network. The input layer serves to distribute signals out of all the inputs of a neuron to all the neurons of the output layer. Neurons of the output layer are the computing units (i.e., they compute

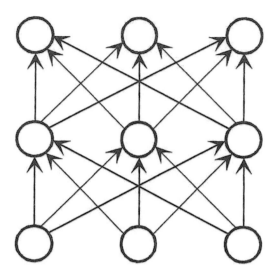

FIGURE 4.9: A feed-forward neural network. Reprint from [8] ©2001 World Scientific.

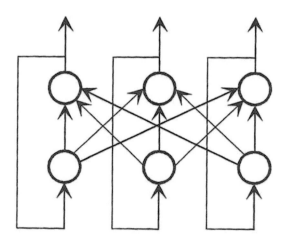

FIGURE 4.10: A feedback neural network. Reprint from [8] ©2001 World Scientific.

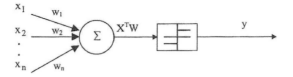

FIGURE 4.11: A simple neuron model. Reprint from [8] ©2001 World Scientific.

their outputs as a function applied to the weighted sum of the input signals). That function can be linear or non-linear. For the linear activation function, the output of the network is determined in the following manner:

$$Y = WX + \theta$$

where W is the weight vector of the network, and X and Y are the input and output vectors correspondingly.

The use of the nonlinear activation function allows an increase in the computational power of the network. For the sigmoid activation function, the network output is determined in the following manner:

$$Y = \frac{1}{1 - \exp^{-XW + \theta}}$$

A multi-layer neural network consists of the input, the output, and the hidden layers. Single-layer networks, which do not have hidden layers, cannot solve complicated problems. The use of the hidden layers allows an increase in the computational power of the network. The outputs of the i-th layer are the functions of the outputs of the $(i - k)$-th (here $k = 1...i - 1$) layers. By choosing an optimal topological structure of a network, an increase in reliability and computational power, as well as a decreased processing time, may be achieved.

4.4.2 Supervised Learning in Neural Networks

The simplest neural network is a perceptron, which is shown in Figure 4.11. Here σ multiplies each input x_i by a weight w_i, $i \in [1, n]$, and sums the weighted inputs. If this sum is greater than the perceptron's threshold, then the output is one; otherwise, it is zero. A perceptron is trained repeatedly by presenting a set of input patterns to its inputs and adjusting the connection weights until the desired output occurs.

Each input pattern of a perceptron can be represented as a vector $X = \{x_1, x_2, ..., x_n\}^T$. The output Y of the perceptron is determined by comparing the weighted sum of the input signals with a threshold value θ: if the weighted sum of the elements of the input vector exceeds θ, the output of the perceptron is one; otherwise, it is zero. Learning is accomplished in the following manner. A pattern X is applied to the input of perceptron, and an output Y is calculated. If the output is correct (i.e., corresponds to the desired one), the weights are not changed. If not, the weights, corresponding to the input connections that cause this incorrect result, are modified to reduce the error. Note that the training must be global; i.e., the perceptron must learn over the entire set of the input patterns, applied to the perceptron either sequentially or randomly. The training method may be generalized by the "delta rule":

- Step 1. Accept the regular input pattern X and calculate output Y for it.

- Step 2.

  - If output Y is correct, go to Step 3.
  - If output Y is incorrect, for each weight w_i, $\triangle w_i = \gamma e x_i$, $w_i(t+1) = w_i(t) + \triangle w_i$, where $e = y^* - y$ is the error for this pattern (y^* is the target output value) and γ is the "learning rate" to regulate the average size of the weight change.

- Step 3. Repeat steps 1–3 until the learning error is at an acceptable level.

Note that this "delta rule" algorithm leads a perceptron to a correct functioning in a finite number of steps. However, we cannot precisely evaluate this number. In certain cases simply trying all possible adjustments of the weights may be sufficient. In addition, it is noted that the representational ability of a perceptron is limited by the condition of the linear separability; there is no way to determine this condition if the dimension of the input vectors is large.

This "delta rule" algorithm can also be used for perceptrons with continuous activation functions. If activation function f is non-linear and differentiable, one may obtain the "delta rule" algorithm in which the correction of the weight coefficients is carried out as follows:

$$\triangle w_i = \gamma(y_i^* - y_i)f'(I)x_i \tag{4.5}$$

$$w_i(t+1) = w_i(t) + \triangle w_i \tag{4.6}$$

where $\triangle w_i$ is the correction associated with the i-th input; $w_i(t)$ is the value of the weight before the adjustment; and $w_i(t+1)$ is the value of the weight after the adjustment.

For training a multi-layer neural network, the least squares procedure must be generalized in order to provide an adequate adjustment for the weight coefficients of the connections, which come to the hidden units. The error back-propagation algorithm [180, 179] is a generalization of the least squares procedure for networks with hidden layers.

When such a generalization is built, the following question occurs: how do we determine the measure of error for the neurons of the hidden layers? This problem is solved by estimating the measure of the error through the error of the units of the subsequent layer. At every step of the learning for each input/output training pair, a first forward pass is performed. This means that the input of a neural network is given by the input vector; as a result, the activation flow passes through the network in the direction from the input layer toward the output. After this process, the states of all the neurons of the network are determined. The output neurons generate the actual output vector, which is compared with the desired output vector, and the learning error is determined. Subsequently, this error is propagated backwards along the network in the direction toward the input layer, updating the values of the weight coefficients.

Thus, the learning process is the consequence of interchanging forward and backward passes; during the forward pass the states of the network units are determined, while during the backward pass the error is propagated and the values of the weights of the connections are updated. That is why this procedure is called the error back-propagation algorithm.

As we mentioned above, increasing the number of layers leads to enhancing the computational power of a network, ultimately, to the possibility of providing much more complex computations. It is shown that a three-layer network is capable of handling convex regions in the input space. Adding a fourth layer may further allow handling non-convex regions [216]. Thus, with the use of four-layer neural networks, practically any computation can be provided. However, adding more layers to a network obviously increases the complexity and learning cost. In addition, with the hidden units in the network, there arises an issue of the optimal number of hidden units in the network.

As is clear from Equation 4.5, for defining the step of updating weight w_i, the determination of the value of the derivative $\partial E/\partial w_i$ is required. This derivative, in turn, is determined through $\partial E/\partial y_j$. In a neural network, we require that the activation function f be differentiable everywhere. For this requirement, the sigmoid function is typically used as an activation function, which has the following derivative:

$$\frac{dy}{dx} = y(1-y)$$

Before the learning process begins, small random values are assigned to all the weight coefficients. It is important that the initial values of the weights not be equal to each other. The above given equation for adjusting weight coefficients is explicitly derived from the gradient descent method

$$\triangle w_i = -\gamma \frac{\partial E}{\partial w_i}$$

where E is the squared error that cumulatively measures the error through all the cases given by the training set. Consequently, all the input training vectors are applied to the network and the measure of the error is obtained. According to this error, corrections are made. This procedure is called the batch version of the error back-propagation algorithm.

There is another approach to updating the weight coefficients. In the case where a single input vector is applied at each time, the current output is generated; given the output error for this single input vector, weight updating is performed. Then the next single input vector is selected and the process is repeated until a convergence is reached. This single weight updating procedure is called the real-time version of the error back-propagation algorithm.

The back-propagation algorithm is conceptually described in Algorithm 5.

Algorithm 5 Back-propagation algorithm

Input: Network topological structure and a set of input-output pairs as the training data
Output: The network connection weights
Method:

1: Initialize the weights in the network (often randomly)
2: **repeat**
3:     **for** Each example s in the training set **do**
4:         O = network output given the input s
5:         T = given output in the training set for s
6:         Calculate error $T - O$ at the output units
7:         Compute $\triangle w_i$ for all the connection weights from the hidden layer to the output layer
8:         Compute $\triangle w_i$ for all the connection weights from the input layer to the hidden layer
9:         Update the weights in the network
10:     **end for**
11: **until** All the examples classified correctly or a stopping criterion satisfied

---

The main downside of the error back-propagation algorithm in Algorithm 5 is the possibility of the network to converge into a local minimum during the back-propagation learning. Another potential issue with this learning algorithm is the low learning speed; i.e., it typically takes many iterations to converge. Complexity-wise, if the network contains Q_1 neurons expected to provide Q_2 output values with Q_3 learning steps, the total learning time is $Q_2 \times Q_1 \times Q_3$. Assuming $Q_1 = Q_2 = Q_3 = Q$, this means that the learning time is proportional to $O(Q^3)$. Using parallel computations decreases this estimation to $O(Q^2)$, as the neurons may work in parallel.

The error back-propagation algorithm is also useful for training the radial

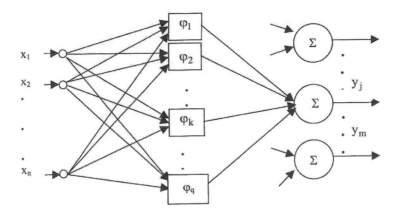

FIGURE 4.12: Multi-input and multi-output structure of RBFNN with a hidden layer. Reprint from [8] ©2001 World Scientific.

based function neural networks (RBFNNs). A series of supervised learning algorithms are employed for training RBFNNs. A RBFNN has a hidden layer of the RBF units, shown in Figure 4.12. This network, consisting of n input and m output neurons, performs the mapping of the n-dimensional input vector X to the m-dimensional output vector Y $(R^n \longrightarrow R^m)$. The node elements φ_k $(k = 1, 2, ..., L)$ of the hidden layer represent the RBFNN transformation for the input signals x_i $(i = 1, 2, ..., n)$. The connections between the input layer and the hidden layer have no corresponding weights. Selection of an appropriate RBF depends on the type of the problems we solve using the RBFNN. As RBF $\varphi_k = \varphi(\|X - a_k\|) = \varphi_k(r)$, the following specific radial functions may be selected:

- $\varphi(r) = r$, a linear radial function;

- $\varphi(r) = r^2$, a quadratic radial function;

- $\varphi(r) = \exp(-r^2/b^2)$, a Gaussian radial function;

- $\varphi(r) = r^2 \log(r)$, a thin-plate-spline radial function;

- $\varphi(r) = \sqrt{(r^2 + b^2)}$, a multi-quadratic radial function.

In Figure 4.12 the response of the k-th hidden unit is given by

$$\varphi(X) = \varphi(\frac{\|X - a_k\|}{b_k^2})$$

where $\varphi(\bullet)$ is a strictly positive radially symmetric function with a unique maximum at the k-th "center" a_k and the function drops off rapidly to zero away from the center; the parameter b_k is the "width" of the receptive field in the input space for unit k.

The supervised learning algorithms for training a RBFNN are based on a training set of N input-output pairs (X^k, Y^k) $(k = 1, 2, ..., N)$ that represent the associations of a given mapping or samples of a continuous multivariate function. RBFNNs have a differentiable nature, providing a differentiation approach to all the parameters for the network training consideration. Therefore, it is possible to use a gradient method to train the parameters of a RBFNN. This method takes the following iteration procedures:

$$a_i(t+1) = a_i(t) + \triangle a_i$$

$$b_i(t+1) = b_i(t) + \triangle b_i$$

$$w_{ij}(t+1) = w_{ij}(t) + \triangle w_{ij}$$

Here $\triangle a_i = -\gamma_a \frac{\partial E}{\partial a_i}$, $\triangle b_i = -\gamma_b \frac{\partial E}{\partial b_i}$, and $\triangle w_{ij} = -\gamma_w \frac{\partial E}{\partial w_{ij}}$, where γ_a, γ_b, and γ_w are small positive constraints and E is a mean-square error between the output of a modeled object (or a function) y_{jr}^* and the corresponding output of the RBFNN y_{jr}:

$$E = \frac{1}{2} \sum_{r=1}^{S} \sum_{j=1}^{M} (y_{jr}^* - y_{jr})^2$$

4.4.3 Reinforcement Learning in Neural Networks

Unlike traditional supervised learning, in which the expected output is given directly and explicitly as part of the training data to guide or "supervise" the learning process, reinforcement learning refers to those learning methods in which a neural network adjusts its behavior based on the given information provided in the training data that indicates an approval or a disapproval of the current behavior of the network. The signal given to the network is called the reinforcement signal, which is sent by a supervisor during the learning process.

Reinforcement learning is developed with an analogy to human learning development. In particular, during children's development, behavior is learned based on the guidance from the environment, including the supervision of an adult. For example, if actions of a child cause non-desirable effects from the environment (e.g., causing pain, or receiving disapproval from an adult), the child will not likely repeat these actions in the future. If, on the other hand, actions of a child cause positive reward from an environment (e.g., praise or approval from an adult), the child will likely repeat these actions in the future.

Due to this analogy, reinforcement learning algorithms are considered biologically plausible. In these algorithms, only a single reinforcement signal is necessary, also called a reward/penalty signal. When the network obtains a reward signal (positive reinforcement), it attempts to repeat the current

behavior in similar circumstances. If a penalty signal (negative or zero reinforcement) arrives, the weight coefficients are modified in order to stay away
from the current behavior.

Using the reinforcement signal for learning is known from the classic automata theory [234, 157]. Consider an automaton, which selects one of the
possible actions to perform based on the corresponding probabilities. When
an action is selected, the environment responds to that action by sending the
positive (reward) or negative (penalty) reinforcement signal to the automaton. Based on this information, the automaton may modify the corresponding
probabilities in order to increase the expectation of the positive reinforcement
signal. This process is carried out repeatedly until the frequency of deriving
the positive reinforcement is at a sufficiently high level.

Consider one of the well-known reinforcement algorithms, the linear reward–
penalty algorithm, L_{R-P} [157]. At first, consider the automaton, which has
the set of selected actions $(a^1, a^2, ..., a^n)$, selected based on the probabilities
$(p^1, p^2, ..., p^n)$. Since during the learning process the probabilities of the action
selection are to be updated, we introduce the additional index, such that
$(p_t^1, p_t^2, ..., p_t^n)$ is the set of the current probabilities in the t-th iteration. At
the t-th learning step action, a_i ($i \in [1, n]$) is selected and the environment
responds to that choice by sending the signal b_t, which has two possible values:
-1 or 1. No additional information is given besides this signal. The automaton
then updates the probabilities p_t^n in accordance with the following rules:

- If the selected actions $a_t = a^i$ and $b_t = 1$, then

$$\begin{cases} p_{t+1}^i = p_t^i + \gamma_1(1 - p_t^i) \\ p_{t+1}^j = (1 - \gamma_1)p_t^j, j \neq i \end{cases} \qquad (4.7)$$

- If $b_t = -1$, then

$$\begin{cases} p_{t+1}^i = (1 - \gamma_2)p_t^i \\ p_{t+1}^j = \frac{\gamma_2}{r-1} + (1 - \gamma_2)p_t^j, j \neq i \end{cases} \qquad (4.8)$$

Here γ_1 and γ_2 are the learning rates, and r is the number of the automata.
Thus, the probability of a successful action is increased to the value proportional to the difference between one and the probability before the learning
step, while probabilities of other actions are decreased. Certainly, γ_1 and γ_2
must be within the range between zero and one.

In the context of neural networks, consider the number of the neurons,
represented in Figure 4.13. Here a^i is considered as the activation of the
i-th neuron. As in the case of an automaton, the learning is accomplished
by repeatedly updating the probabilities for neuron firing. However, in this
case, individual neurons do not compose a global network. Consequently, they
lose the computational power derived by using connections, and thus cannot
accomplish associations and classifications.

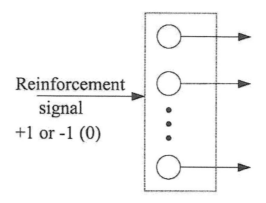

FIGURE 4.13: Neurons for L_{R-P} algorithm. Reprint from [8] ©2001 World
Scientific.

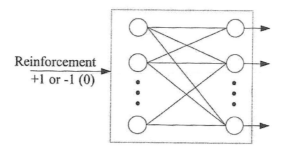

FIGURE 4.14: Network for A_{R-P} algorithm. Reprint from [8] ©2001 World
Scientific.

Taking into account this issue in [16], the enhanced L_{R-P} algorithm is
developed, which is called associative reinforcement learning. A network is
considered that contains both output and input neurons as illustrated in Figure 4.14. The input vectors are connected to the input units in order to
accomplish the classification. The network is trained to perform the correct
classification with the use of a reinforcement signal, derived from the training
data. In order to effectively reward the correct network behavior, the information on the relation between the input vectors X_k and the output vectors
Y_k must be given. One way to store such information is to use the array
$d(X_k, Y_k)$.

Assume that there are two neurons in the output layer of the network. This
means that the input vectors correspond to two different classes. When a particular input vector X is connected to the input units, the classification error

is minimized if the neuron y_1 fires under the condition $P(y_1|X) > P(y_2|X)$ or if the neuron y_2 fires under the condition $P(y_2|X) > P(y_1|X)$. The challenge is how to determine the given conditional probabilities.

Barto and Anandan [16] have suggested using the vector Θ, which approximates those probabilities:

$$\Theta X = P(y_1|X) - P(y_2|X)$$

Thus, if the condition $\Theta X > 0$ is satisfied, the output neuron y_1 fires; otherwise, the second output neuron fires. The vector Θ is adjusted during the learning process. Besides the above firing rule, the class label Z of the input vector is introduced. This class label is equal to 1 if X is in the class, which corresponds to y_1; otherwise, $Z = -1$. It is shown that minimizing the mathematical expectation of $(\Theta X - Z)^2$ causes minimizing the classification error.

To achieve this goal, the Robinson-Monro algorithm [121] may be used. The partial derivative of error E on Θ is defined as follows:

$$\frac{\partial E}{\partial \Theta} = 2(\Theta X - 2)X \qquad (4.9)$$

The equation below is used to adjust the vector Θ during the learning process:

$$\Theta_{t+1} = \Theta_t - \gamma_t(\Theta_t - Z_t)X_t \qquad (4.10)$$

where γ_t is a constant for each t; these constants have different values at different steps of the learning process. Their values decrease during the learning process and influence the convergence of the algorithm.

Components of vector Θ can be considered as the weights of the connections, which connect the input neurons to one of the output neurons. These output neurons are activated (fired) when their total input exceeds zero; otherwise, they are not activated.

For developing an associative reward/penalty algorithm, A_{R-P}, the randomness elements are used. In [16], it is assumed that each output neuron can be in two states: 1 and -1. The activation rule takes the following form:

$$y_t = \begin{cases} 1 \text{ if } X_t + \xi_t > 0, \\ -1 \text{ otherwise.} \end{cases}$$

Here ξ_t is a random variable with a known distribution. When X_t and Θ_t are given, the mathematical expectation $E(y_t|\Theta_t, X_t)$ is known, too. The mathematical definition for Θ updating stays similar to that in the Robinson-Monro algorithm (i.e., Equation 4.9).

To distinguish the case of the positive reinforcement ($b = 1$) from the negative one ($b = -1$), the coefficient λ is introduced. In the reward case, we have:

$$\Theta_{t+1} = \Theta_t - \gamma_t(E(y_t|\Theta_t, X_t) - b_t y_t)X_t$$

In the case of penalty:

$$\Theta_{t+1} = \Theta_t - \lambda\gamma_t(E(y_t|\Theta_t, X_t) - b_ty_t)X_t$$

When $\lambda = 0$, the above algorithm is called the associative reward in action algorithm. It is shown that if:

1. the input vectors are linearly independent;

2. the appearance of each input vector has a finite probability;

3. the distribution of the random variable is continuous and monotonous; and

4. the sequence of γ_k satisfies certain requirements, providing minimizing γ_k to zero as k increases,

then the weight vector converges.

The main downside of reinforcement learning is that it is not efficient in solving large problems. In large networks, it is difficult to adjust the behavior, based only on the single global reward/penalty signal. Another problem is that the ascent toward increasing the reinforcement signal expectation can lead to a local optimum. When the network moves closer to that optimum, it derives less information about other possible solutions.

4.5   Genetic Algorithms

While reinforcement learning algorithms are developed under the analogy to human behavior development, genetic algorithms are developed under the analogy to natural process development. The known algorithms in this paradigm developed in analogy to natural process development include evolutionary programming, genetic algorithms, evolution strategies, and simulated annealing. Furthermore, recent research has led to the generalization of genetic algorithms to *evolution programs*. Classic genetic algorithms operate on fixed-length binary strings, which do not need to be the case for evolution programs. Also, evolution programs usually incorporate a variety of "genetic" operators, whereas classic genetic algorithms use the binary crossover and mutation operators only.

4.5.1   Genetic Algorithms in a Nutshell

The beginning of genetic algorithms can be traced back to the early 1950s, when several biologists used computers for simulations of biological systems.

However, the work done in late 1960s and early 1970s led to the formal development of the genetic algorithms, as they are known today.

Genetic algorithms (GAs) are global optimization algorithms based on the mechanics of natural selection and natural genetics. GAs have a number of specific peculiarities by which they differ from the other methods of optimization. These include:

1. GAs employ only an objective function, not the derivative function or some other information on the object. It is convenient in case that the function is neither differentiable nor discrete.

2. GAs employ a parallel multi-point search strategy by maintaining a population of the potential solutions, which provide wide information on the function behavior and exclude the possibility of sticking to a local extremum of the function, while the traditional search methods are typically unable to cope with this problem.

3. GAs use the probability-transitive rules instead of the deterministic rules. Besides, GAs are very simple for computer implementation.

Genetic algorithms use a vocabulary borrowed from natural genetics. A candidate solution is called an *individual*. Quite often this individual is also called a *string* or a *chromosome*. This might be a bit misleading: each cell of every organism of a given species carries a certain number of chromosomes; however, we talk about only one individual chromosome. Chromosomes are made of units — genes arranged in a linear succession; every gene controls the inheritance of one of several characteristics.

Each gene can assume a finite number of values. In a binary representation, a chromosome is a vector, consisting of the bits in succession, i.e., the succession of zeroes and ones. A set of chromosomes makes a population. The number of the chromosomes in a population defines the population size. The genetic algorithms evaluate a population and generate a new one iteratively, with each successive population referred to as a generation. The population undergoes a simulated evolution: at each generation the relatively "good" solutions reproduce, while the relatively "bad" solutions die. To distinguish between different solutions we use an objective (evaluation) function, which plays the role of an environment. Quite often the objective function is also called the fitness function.

The structure of a simple genetic algorithm is the same as the structure of any evolution program. During iteration t, a genetic algorithm maintains a population of the potential solutions (chromosomes, vectors), $G(t) = \{x_1^t, ..., x_n^t\}$. Each solution x_i^t is evaluated to give some measure of its "fitness". Then, a new population (iteration $t+1$) is formed by selecting the more fit individuals. Some members of this new population undergo reproduction by means of crossover and mutation, to form new solutions. The crossover combines the features of two parent chromosomes to form two similar offsprings by swapping the corresponding segments of the parents. For example,

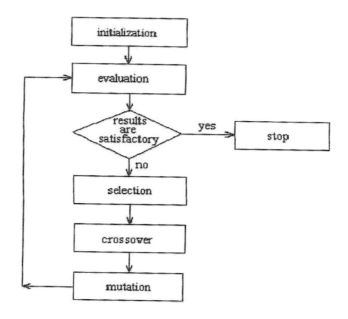

FIGURE 4.15: The structure of a simple genetic algorithm. Reprint from [8] ©2001 World Scientific.

if the parents are represented by five-dimensional vectors $(a_1, b_1, c_1, d_1, e_1)$ and $(a_2, b_2, c_2, d_2, e_2)$, a possible crossover may generate the offsprings $(a_1, b_1, c_2, d_2, e_2)$ and $(a_2, b_2, c_1, d_1, e_1)$.

The mutation is defined to arbitrarily alert one or more genes of a selected chromosome by a random change, with a probability equal to the mutation rate. For a particular problem, a genetic algorithm is presented in Figure 4.15.

We discuss the actions of a genetic algorithm for a simple parameter optimization problem. Suppose that we wish to maximize a function of k variables $f(x_1, ..., x_k) : R^k \longrightarrow R$. If the optimization problem is to minimize the function f, this is equivalent to maximizing the function g, where $g = -f$, i.e.,

$$\min\{f(x)\} = \max\{g(x)\} = \max\{-f(x)\}$$

Suppose further that each variable x_i can take values from a domain $D_i = [a_i, b_i] \subseteq R$ and $f(x_1, ..., x_k) > 0$ for all $x_i \in D_i$. We wish to optimize the function f with a required precision: suppose that six decimal digits for the variables' values are desirable.

It is clear that to achieve such precision each domain D_i should be cut into $(b_i - a_i)10^6$ equal-size units. Let us denote by m_i the smallest number of the bits such that $(b_i - a_i)10^6 \leq 2^{m_i} - 1$. Then, a representation having such a variable x_i coded as a binary string of length m_i clearly satisfies the precision requirement. Additionally, the following formula interprets each such string:

$$x_i = a_i + decimal(1001...001_2) \times \frac{b_i - a_i}{2^{m_i} - 1}$$

where decimal($string_2$) represents the decimal value of that binary string.

Now each chromosome (as a potential solution) is represented by a binary string of length $m = \sum_{i=1}^{k} m_i$; the first m_1 bits map into a value from the range $[a_1, b_1]$, the next group of m_2 bits map into a value from the range $[a_2, b_2]$, and so on; the last group of m_k bits map into a value from the range $[a_k, b_k]$.

To initialize a population, we can simply set some p number of chromosomes randomly in a bitwise fashion. However, if we have a priori knowledge about the distribution of the potential optima, we may use such information in arranging the set of the initial (potential) solutions. The rest of the algorithm is straightforward: in each generation we evaluate each chromosome (using the function f on the decoded sequences of the variables), select a new population with respect to the probability distribution based on the fitness values, and recombine the chromosomes in the new population by mutation and crossover operators. After a number of generations, when no further improvement is observed, the best chromosome represents a (possibly the global) optimal solution. Often we stop the algorithm after a fixed number of iterations, depending upon the speed and resource criteria.

For the selection process (selection of a new population with respect to the probability distribution based on the fitness values), we must implement the following actions at first:

- Compute the fitness value u_i for each chromosome v_i, $(i = [1, p])$

- Find the total fitness of the population $F = \sum_{i=1}^{p} u_i$

- Compute the probability of a selection p_s^i for each chromosome $v_i(i = [1, p])$: $p_s^i = u_i/F$

- Compute a cumulative probability p_{cum}^i for each chromosome $v_i(i = [1, p])$: $p_{cum}^i = \sum_{j=1}^{i} p_s^j$

The selection process is implemented p times; each time we select a single chromosome for a new population in the following way:

- Generate a random (float) number r from the range [0,1].

- If $r < p_{cum}^1$, then select the first chromosome (v_1); otherwise, select the i-th chromosome v_i ($2 \leq i \leq p$ such that $p_{cum}^{i-1} < r < p_{cum}^i$).

Obviously, some chromosomes may be selected more than once; the best chromosomes generate more copies, the average ones remain, and the worst die off. Now we are ready to apply the first combination operator, crossover, to the individuals in the new population. One of the parameters of a genetic system is the probability of crossover p_c. The probability gives us the expected number $p_c \times p$ of the chromosomes, which undergo the crossover operation. We proceed in the following way. For each chromosome in the new population:

- Generate a random (float) number r from the range $[0, 1]$;

- if $r < p_c$, select the given chromosome for crossover.

Now we mate the selected chromosomes randomly: for each pair of the coupled chromosomes, we generate a random integer number *pos* from the range $[1, m - 1]$ (m is the total length — the number of the bits in a chromosome). The number *pos* indicates the position of the crossing point. Two chromosomes

$$(b_1, b_2, ..., b_{pos}, b_{pos+1}, ..., b_m)$$

and

$$(c_1, c_2, ..., c_{pos}, c_{pos+1}, ..., c_m)$$

are replaced by a pair of their offsprings:

$$(b_1, b_2, ..., b_{pos}, c_{pos+1}, .., c_m)$$

and

$$(c_1, c_2, ..., c_{pos}, b_{pos+1}, ..., b_m)$$

The intuition behind this application of the crossover operator is the information exchange between the different potential solutions.

The next combination operator, mutation, is performed on a bit-by-bit basis. Another parameter of the genetic system, the probability of mutation p_m, gives us the expected number of the bits $p_m \times m \times p$. Every bit (in all chromosomes in the whole population) has an equal chance to undergo a mutation, i.e., changing from 0 to 1 or vice versa. Thus, we proceed in the following way. For each chromosome in the current (i.e., after the crossover) population and for each bit within the chromosome:

- Generate a random (float) number r from the range $[0, 1]$;

- if $r < p_m$, mutate the bit.

The intuition behind this application of the mutation operator is the introduction of an extra variation into the population.

Following the selection, crossover, and mutation, the new population is ready for its next stage of the development. This stage of the development is used to build the probability distribution (for the next selection process).

The rest of the stage of the development (or the evolution) is just a cyclic repetition of the above steps.

However, as may frequently occur, in the earlier generations the fitness values of some chromosomes may be better than the values of the best chromosomes after a finite number of the generations. It is relatively easy to keep track of the best individual in the evolution process. It is customary (in the genetic algorithms' implementation) to store "the best ever" individual at a separate location; in that way, the algorithm would report the best value found during the whole process (as opposed to the best value in the final population).

It is necessary to note that a classic genetic algorithm may employ the roulette wheel method for the selection, which is a stochastic version of the survival-of-the-fittest mechanism. In this method of the selection, candidate strings from the current generation $G(t)$ are selected to survive in the next generation $G(t + 1)$ by designing a roulette wheel where each string in the population is represented on the wheel in proportion to its fitness value. Thus, those strings which have a higher fitness value are given a larger share of the wheel, while those with a lower fitness value are given a relatively smaller portion of the roulette wheel. Finally, selections are made by spinning the roulette wheel p times and candidates are accepted for those strings which are indicated at the completion of the spin.

4.5.2   Comparison of Conventional and Genetic Algorithms for an Extremum Search

Genetic algorithms of an extremum search differ to a great extent from the conventional methods of optimization.

Let us find the maximum of the function $f_1(x)$ illustrated in Figure 4.16 using the gradient method. This method helps solve the problem rather quickly, starting from an initial point and gradually approaching the top.

However, if we use the same method to find the global optimum of the function $f_2(x)$ illustrated in Figure 4.17, we would be stuck in a local optimum, in the neighborhood of which the initial approximation was chosen. Nevertheless, a genetic algorithm operated by a population of the points may approach the global optimum without any risk of being stuck in a local optimum.

On the other hand, for the same genetic algorithm there are always different variations proposed by different authors, with the aim to facilitate the algorithm and to make it more effective. For this purpose it is also suggested to use hybrid algorithms, combining genetic algorithms with conventional learning algorithms, such as the gradient descent, hill-climbing, and coordinate methods. These hybrid methods have shown a high efficiency for a certain class of the problems.

One of the essential advantages of genetic algorithms is the *black box* principle. That is, it is enough to give the input data X and to obtain the output Y without necessarily knowing the actual "function" expected. In genetic

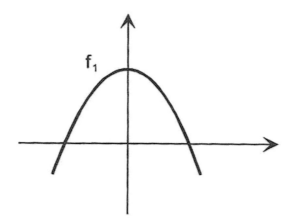

FIGURE 4.16: Graph of the function f_1. Reprint from [8] ©2001 World Scientific.

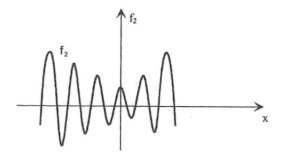

FIGURE 4.17: Graph of the function f_2. Reprint from [8] ©2001 World Scientific.

algorithms, it is not always required to code variables from the decimal representation to the binary representation, or vice versa, in each generation. Also, there is no need to compute the derivative of a function f or any other additional information about the function. In comparison, in order to use the traditional search methods, differentiability and continuity of an analytical form of a function are usually required. All these advantages make genetic algorithms attractive in solving complex problems.

Let us discuss three algorithms for a more specific comparison. They are the hill-climbing, simulated annealing, and genetic algorithms. We apply these algorithms to a simple optimization problem. This comparison underlines the uniqueness of the genetic algorithm approach. The search space in this application is a set of binary strings v of length 30. The objective function f to be maximized is given as

$$f(v) = |11 \times one(v) - 150|$$

where the function $one(v)$ returns the number of 1s in the string v.

The function f is linear and does not provide any challenge as an optimization task. We use it only to illustrate the difference in performances of these three algorithms. However, the interesting point of the function f is that it has one global maximum for

$$v_g = (111111111111111111111111111111)$$

$f(v_g) = |11 \times 30 - 150| = 180$, and one local maximum for

$$v_l = (000000000000000000000000000000)$$

$f(v_l) = |11 \times 0 - 150| = 150$.

There are several versions of the hill-climbing algorithm. They differ in the way a new string is selected for comparison with the current string. Algorithm 6 lists a simple version of the hill-climbing algorithm (the steepest ascent hill-climbing).

Initially, all the 30 neighbors are considered, and the v_n which returns the largest $f(v_n)$ is selected to compete with the current string v_c. If $f(v_c) < f(v_n)$, then the new string becomes the current string. Otherwise, no local improvement is possible — the algorithm has reached the optimum. In this case, the next iteration ($t \longleftarrow t + 1$) of the algorithm is executed with another string selected at random.

It is interesting to note that the starting string (randomly selected) determines the success or failure of the single iteration of the above hill-climbing algorithm (i.e., return of the global or local optimum). It is clear that if the starting string has thirteen 1s or less, the algorithm will always terminate in the local optimum (failure). The reason is that a string with thirteen 1s returns a value 7 of the objective function, and any single-step improvement toward the global optimum, i.e., to increase the number of 1s to fourteen,

---

Algorithm 6 The steepest ascent hill-climbing algorithm

Input: A set of strings
Output: The best string corresponding to the maximum objective function value
Method:

  1: Initialize $t \longleftarrow 0$
  2: **repeat**
  3:     $local \longleftarrow FALSE$
  4:     Select a current string v_c at random
  5:     Evaluate v_c
  6:     **repeat**
  7:         Select 30 new strings in the neighborhood of v_c by flipping single bits of v_c
  8:         Select the string v_n from the set of the new strings with the largest value of the objective function f
  9:         **if** $f(v_c) < f(v_n)$ **then**
10:            $v_c \longleftarrow v_n$
11:         **else**
12:            $local \longleftarrow TRUE$
13:         **end if**
14:     **until** $local = \text{TRUE}$
15:     $t \longleftarrow t + 1$
16: **until** $t == MAX$

---

decreases the value of the objective function to 4. On the other hand, any decrease in the number of 1s would increase the value of the function — a string with twelve 1s yields a value of 18, a string with eleven 1s yields a value of 29, etc. This would push the search in the *wrong* direction, toward the local maximum. For the problem with multiple local optima, the chance of hitting the global optimum (in a single iteration) is even weaker.

The algorithm of the simulated annealing is given in Algorithm 7. The Boltzmann probability is

$$p = \exp((f(v_n) - f(v_c))/T)$$

Algorithm 7 Simulated annealing algorithm

Input: A set of strings
Output: The best string corresponding to the maximum objective function value
Method:

1: Initialize $t \longleftarrow 0$
2: Initialize temperature T
3: Select a current string v_c at random
4: Evaluate v_c
5: **repeat**
6:    **repeat**
7:       Select a new string v_n in the neighborhood of v_c by flipping a single bit of v_c
8:       **if** $f(v_c) < f(v_n)$ **then**
9:          $v_c \longleftarrow v_n$
10:       **else**
11:          **if** $random[0,1] < \exp\{(f(v_n) - f(v_c))/T\}$ **then**
12:             $v_c \longleftarrow v_n$
13:          **end if**
14:       **end if**
15:    **until** Termination condition met
16:    $T \longleftarrow g(T,t)$
17:    $t \longleftarrow t+1$
18: **until** Stop criterion met

The function $random[0,1]$ returns a random number from the range $[0,1]$. The termination condition checks whether the *thermal equilibrium* has been reached, i.e., whether the probability distribution of the selected new strings approaches the expected distribution. However, in certain implementations, this loop is executed just k times (k is an additional parameter of the Boltzmann method).

The temperature T is lowered in steps ($g(T, t) < T$ for all t). The algorithm terminates for a small value of T — the stop criterion checks whether the system is *frozen*, i.e., virtually no changes are accepted anymore.

Since the simulated annealing algorithm can escape local optima, let us consider a string

$$v_s = (111000000100110111001010100000)$$

with twelve 1s, which evaluates to $f(v_s) = |11 \times 12 - 150| = 18$. For v_s as the starting string, the hill-climbing algorithm (as discussed above) would approach the local maximum

$$v_h = (000000000000000000000000000000)$$

since any string with thirteen 1s (i.e., *toward* the global optimum) evaluates to 7 (less than 18). On the other hand, the simulated annealing algorithm would accept a string with thirteen 1s as a new current string with a probability which, for a certain temperature such as $T = 20$, gives

$$p = e^{-\frac{11}{20}} = 0.57695$$

i.e., the chance of acceptance is better than 50%.

Generic algorithms maintain a population of strings. Two relatively poor strings

$$v_p = (111110000000110111001110100000)$$

and

$$v_q = (000000000001101110010101111111)$$

each of which evaluates to 16, can produce much better offsprings (if the crossover point falls anywhere between the 5th and the 12th position):

$$v_r = (111110000001101110010101111111)$$

The new offspring v_r evaluates to

$$f(v_r) = |11 \times 19 - 150| = 59$$

The above comparison analysis through the given simple example illustrates the advantages of the genetic algorithms over the traditional optimization methods.

4.6 Summary

In this chapter we have reviewed soft computing techniques as an approach to multimedia data mining. Specifically, we have studied three different paradigms of soft computing: fuzzy sets and logic, neural networks, and genetic algorithms. These paradigms can be applied to various machine learning and optimization tasks in general and to multimedia data mining in particular. It is noted that these different paradigms of soft computing complement each other, rather than compete with each other. It becomes clear that fuzzy sets and logic, neural networks, and genetic algorithms are more effective when used in combinations. Some combination examples include:

- neural networks + fuzzy logic (Neuro-Fuzzy);

- Fuzzy logic + genetic algorithms;

- Neural network + genetic algorithms;

- Fuzzy logic + neural networks + genetic algorithms;

- Combinations of the other paradigms of soft computing are also possible.

Various applications of these soft computing techniques are further discussed in developing specific solutions to different multimedia data mining problems in the chapters in Part III of this book.

Part III

Multimedia Data Mining Application Examples

Chapter 5

Image Database Modeling – Semantic Repository Training

5.1 Introduction

This chapter serves as an example to investigate content based image database mining and retrieval, focusing on developing a classification-oriented methodology to address semantics-intensive image retrieval. In this specific approach, with Self Organization Map (SOM) based image feature grouping, a visual dictionary is created for color, texture, and shape feature attributes, respectively. Labeling each training image with the keywords in the visual dictionary, a classification tree is built. Based on the statistical properties of the feature space, we define a structure, called an α-*semantics graph*, to discover the hidden semantic relationships among the semantic repositories embodied in the image database. With the α-semantics graph, each semantic repository is modeled as a unique fuzzy set to explicitly address the semantic uncertainty existing and overlapping among the repositories in the feature space. An algorithm using classification accuracy measures is developed to combine the built classification tree with the fuzzy set modeling method to deliver semantically relevant image retrieval for a given query image. The experimental evaluations have demonstrated that the proposed approach models the semantic relationships effectively and outperforms a state-of-the-art content based image mining system in the literature in both effectiveness and efficiency.

The rest of the chapter is organized as follows. Section 5.2 introduces the background of developing this semantic repository training approach to image classification. 5.3 briefly describes the previous work. In Section 5.4, we present the image feature extraction method as well as the creation of visual dictionaries for each feature attribute. In Section 5.5 we introduce the concept of the α-semantics graph and show how to model the fuzzy semantics of each semantic repository from the α-semantics graph. Section 5.6 describes the algorithm we have developed to combine the classification tree built and the fuzzy semantics model constructed for the semantics-intensive image mining and retrieval. Section 5.7 documents the experimental results and evaluations. Finally, the chapter is concluded in Section 5.8.

5.2   Background

Large collections of images have become popular in many multimedia data mining applications, from photo collections to Web pages or even video databases. To effectively index and/or mine them is a challenge which is the focus of many research projects (for instance, the classic IBM's QBIC [80]). Almost all of these systems generate low-level image features such as color, texture, shape, and motion for image mining and retrieval. This is partly because low-level features can be computed automatically and efficiently. The semantics of the images, which users are mostly interested in, however, are seldom captured by the low-level features. On the other hand, there is no effective method yet to automatically generate good semantic features of an image. One common compromise is to obtain the semantic information through manual annotation. Since visual data contain rich information and manual annotation is subjective and ambiguous, it is difficult to capture the semantic content of an image using words precisely and completely, not to mention the tedious and labor-intensive work involved.

One compromise to this problem is to organize the image collection in a meaningful manner using image classification. Image classification is the task of classifying images into (semantic) categories based on the available training data. This categorization of images into classes can be helpful both in the semantic organizations of image collections and in obtaining automatic annotations of the images. The classification of natural imagery is difficult in general due to the fact that images from the same semantic class may have large variations and, at the same time, images from different semantic classes might share a common background. These issues limit and further complicate the applicability of the image classification or categorization approaches proposed recently in the literature.

A common approach to image classification or categorization typically addresses the following four issues: (i) image features — how to represent an image; (ii) organization of the feature data — how to organize the data; (iii) classifier — how to classify an image; and (iv) semantics modeling — how to address the relationships between the semantic classes.

In this chapter, we describe and present a new classification oriented methodology to image mining and retrieval. We assume that a set of training images with known class labels is available. Multiple features (color, texture, and shape) are extracted for each image in the collection and are grouped to create visual dictionaries. Using the visual dictionaries for the training images, a classification tree is constructed. Once the classification tree is obtained, any new image can be classified easily. On the other hand, to model the semantic relationships between the image repositories, a representation called an α-semantics graph is generated based on the defined semantics correlations for each semantic repository pairs. Based on the α-semantics graph, each semantic repository is modeled as a unique fuzzy set to explicitly address the semantic uncertainty and the semantic overlap between the semantic repositories in the feature space. A retrieval algorithm is developed based on the classification tree and the fuzzy semantics model for the semantics-relevant image mining and retrieval.

We have evaluated this method on 96 fairly representative classes of the COREL image database [2]. These image classes are, for instance, fashion models, aviation, cats and kittens, elephants, tigers and whales, flowers, night scenes, spectacular waterfalls, castles around the world, and rivers. These images contain a wide range of content (scenery, animals, objects, etc.). Comparing this method with the nearest-neighbors technique [69], the results indicate that this method is able to perform consistently better than the well-known nearest-neighbors algorithm with a shorter response time.

5.3   Related Work

Very few studies have considered data classification on the basis of image features in the context of image mining and retrieval. In the general context of data mining and information retrieval, the majority of the related work has been concerned with handling textual information [131, 41]. Not much work has been done on how to represent imagery (i.e., image features) and how to organize the features. With the high popularity and increasing volume of images in centralized and distributed environments, it is evident that the repository selection methods based on textual description is not suitable for visual queries, where the user's queries may be unanticipated and referring to unextracted image content. In the rest of this section, we review some of the previous work in automatic classification based image mining and retrieval.

Yu and Wolf presented a one-dimensional *Hidden Markov Model* (HMM) for indoor/outdoor scene classification [229]. An image is first divided into horizontal (or vertical) segments, and each segment is further divided into blocks. Color histograms of blocks are used to train HMMs for a preset standard set of clusters, such as a cluster of sky, tree, and river, and a cluster of sky, tree, and grass. Maximum likelihood classifiers are then used to classify an image as indoor or outdoor. The overall performance of classification depends on the standard set of clusters which describe the indoor scene and outdoor scene. In general, it is difficult to enumerate an exhaustive set to cover a general case such as indoor/outdoor. The *configural recognition* scheme proposed by Lipson et al [140] is also a knowledge-based scene classification method. A model template, which encodes the common global scene configuration structure using qualitative measurements, is handcrafted for each category. An image is then classified to a category whose model template best matches the image by deformable template matching (which requires intensive computation, despite the fact that the images are subsampled to low resolutions) — the nearest neighbor classification. To avoid the drawbacks of manual templates, a learning scheme that automatically constructs a scene template from a few examples is proposed by [171]. The learning scheme was tested on two scene classes and suggested promising results.

One early work for resource selection in distributed visual information systems was reported by Chang et al [42]. The method proposed was based on a meta database at a query distribution server. The meta database records a summary of the visual content of the images in each repository through image templates and statistical features. The selection of the database is driven by searching the meta database using a nearest-neighbor ranking algorithm that uses query similarity to a template and the features of the database associated with the template. Another approach [110] proposes a new scheme for automatic hierarchical image classification. Using banded color correlograms, the approach models the features using singular value decomposition (SVD) [56] and constructs a classification tree. An interesting point of this approach is the use of correlograms. The results suggest that correlograms have more latent semantic structures than histograms. The technique used extracts a certain form of knowledge to classify images. Using a noise-tolerant SVD description, the image is classified in the training data using the nearest neighbor with the first neighbor dropped. Based on the performance of this classification, the repositories are partitioned into subrepositories, and the interclass disassociation is minimized. This is accomplished through using normalized cuts. In this scheme, the content representation is weak (only using color and some kind of spatial information), and the overlap among semantic repositories in the feature space is not addressed.

Chapelle et al. [43] used a trained Support Vector Machine (SVM) to perform image classification. A color histogram was computed to be the feature for each image and several "one against the others" SVM classifiers [20] were combined to determine the class a given image was designated to. Their results show that SVM can generalize well compared with other methods. However, their method cannot provide quantitative descriptions for the relationships among classes in the database due to the "hard" classification nature of SVM (one image either belongs to one class or not), which limits its effectiveness to image mining and retrieval. More recently, Djeraba [63] proposed a method for classification based image mining and retrieval. The method exploited the associations among color and texture features and used such associations to discriminate image repositories. The best associations were selected on the basis of confidence measures. Reasonably accurate retrieval and mining results were reported for this method, and the author argued that content- and knowledge-based mining and retrieval were more efficient than the approaches based on content exclusively.

In the general context of content-based image mining and retrieval, although many visual information systems have been developed [114, 166], except for a few cases such as those reviewed above, none of these systems ever considers knowledge extracted from image repositories in the mining process. The semantics-relevant image selection methodology discussed in this chapter offers a new approach to discover hidden relationships between semantic repositories so as to leverage the image classification for better mining accuracy.

5.4   Image Features and Visual Dictionaries

To capture as much content as possible to describe and distinguish images, we extract multiple semantics-related features as image signatures. Specifically, the proposed framework incorporates color, texture, and shape features to form a feature vector for each image in the database. Since image features $f \in \mathbb{R}^n$, it is necessary to perform regularization on the feature set such that the visual data can be indexed efficiently. In the proposed approach, we create a visual dictionary for each feature attribute to achieve this objective.

5.4.1   Image Features

The color feature is represented as a color histogram based on the CIELab space [38] due to its desired property of the perceptual color difference proportional to the numerical difference in the CIELab space. The CIELab space is quantized into 96 bins (6 for L, 4 for a, and 4 for b) to reduce the computational intensity. Thus, a 96-dimensional feature vector C is obtained for each image as a color feature representation.

To extract texture information of an image, we apply a set of Gabor filters [145], which are shown to be effective for image mining and retrieval [143], to the image to measure the response. The Gabor filters are one kind of two-dimensional wavelets. The discretization of a two-dimensional wavelet applied on an image is given by

$$W_{mlpq} = \iint I(x,y)\psi_{ml}(x - p\triangle x, y - q\triangle y)dxdy \qquad (5.1)$$

where I denotes the processed image; $\triangle x$, $\triangle y$ denote the spatial sampling rectangle; p, q are image positions; and m, l specify the scale and orientation of the wavelets, respectively. The base function $\psi_{ml}(x,y)$ is given by

$$\psi_{ml}(x,y) = a^{-m}\psi(\widetilde{x}, \widetilde{y}) \qquad (5.2)$$

where

$$\widetilde{x} = a^{-m}(x\cos\theta + y\sin\theta)$$
$$\widetilde{y} = a^{-m}(-x\sin\theta + y\cos\theta)$$

denote a dilation of the mother wavelet (x, y) by a^{-m}, where a is the scale parameter, and a rotation by $\theta = l \times \triangle\theta$, where $\triangle\theta = 2\pi/L$ is the orientation sampling period.

In the frequency domain, with the following Gabor function as the mother wavelet, we use this family of wavelets as the filter bank:

$$\Psi(u, v) = \exp\left\{-2\pi^2(\sigma_x^2 u^2 + \sigma_y^2 v^2)\right\} \otimes \delta(u - W)$$
$$= \exp\left\{-2\pi^2(\sigma_x^2(u - W)^2 + \sigma_y^2 v^2)\right\}$$
$$= \exp\left\{-\frac{1}{2}\left(\frac{(u - W)^2}{\sigma_u^2} + \frac{v^2}{\sigma_v^2}\right)\right\} \tag{5.3}$$

where $\otimes$ is a convolution symbol, $\delta(.)$ is the impulse function, $\sigma_u = (2\pi\sigma_x)^{-1}$, and $\sigma_v = (2\pi\sigma_y)^{-1}$. The constant W determines the frequency bandwidth of the filters.

Applying the Gabor filter bank to an image results, for every image pixel (p, q), in an M (the number of scales in the filter bank) by L array of responses to the filter bank. We only need to retain the magnitudes of the responses:

$$F_{mlpq} = |W_{mlpq}| \quad m = 0, \ldots, M - 1, \; l = 0, \ldots L - 1 \tag{5.4}$$

Hence, a texture feature is represented as a vector, with each element of the vector corresponding to the energy in a specified scale and orientation sub-band w.r.t. a Gabor filter. In the implementation, a Gabor filter bank of 6 orientations and 4 scales is performed for each image in the database, resulting in a 48-dimensional feature vector T (24 means and 24 standard deviations for $|W_{ml}|$) for the texture representation.

The edge map is used with the water-filling algorithm [253] to describe the shape information for each image due to its effectiveness and efficiency for image mining and retrieval [154]. An 18-dimensional shape feature vector, S, is obtained by generating edge maps for each image in the database.

Figure 5.1 shows visualized illustrations of the extracted color, texture, and shape features for an example image. These features describe the content of images and are used to index the images.

5.4.2 Visual Dictionary

The creation of the visual dictionary is a fundamental preprocessing step necessary to index features. It is not possible to build a valid classification tree without the preprocessing step in which similar features are grouped. The centers of the feature groups constitute the visual dictionary. Without the visual dictionary, we would have to consider all feature values of all images, resulting in a situation where very few feature values are shared by images, which makes it impossible to discriminate repositories.

For each feature attribute (color, texture, and shape), we create a visual dictionary, respectively, using the Self Organization Map (SOM) [130] approach. SOM is ideal for the problem, as it can project high-dimensional

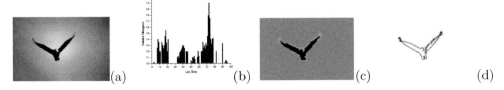

(a) (b) (c) (d)

FIGURE 5.1: An example image and its corresponding color, texture, and shape feature maps. (a) The original image. (b) The CIELab color histogram. (c) The texture map. (d) The edge map. Reprint from [244] ©2004 ACM Press.

feature vectors to a 2-dimensional plane, mapping similar features together while separating different features at the same time.

A procedure is designed to create "keywords" in the dictionary. The procedure follows 4 steps:

1. Performing the Batch SOM learning [130] algorithm on the region feature set to obtain the visualized model (node status) displayed in a 2-dimensional plane map;

2. Considering each node as a "pixel" in the 2-dimensional plane such that the map becomes a binary image, with the value of each pixel i defined as follows:

$$p(i) = \begin{cases} 0 & \text{if } count(i) \geq t \\ 255 & \text{else} \end{cases}$$

   where $count(i)$ is the number of features mapped to the node i and the constant t is a preset threshold. The pixel value 255 denotes objects, while pixel value 0 denotes the background;

3. Performing the morphological erosion operation [38] on the resulting binary image p to make sparse connected objects in the binary image p disjointed. The size of the erosion mask is determined to be the minimum that makes two sparse connected objects separated;

4. With the connected component labeling [38], we assign each separated object a unique ID, a "keyword". For each "keyword", the mean of all the features is determined and stored. All "keywords" constitute the visual dictionary for the corresponding feature attribute.

In this way, the number of "keywords" is adaptively determined and the similarity-based feature grouping is achieved. Applying this procedure to each feature attribute, a visual dictionary is created for each one. Figure 5.2 shows the generation of the visual dictionary. Each entry in a dictionary is one "keyword" representing the similar features. The experiments show that the visual dictionary created captures the clustering characteristics in the feature set very well.

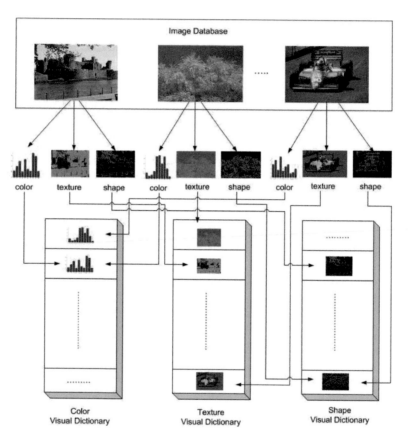

FIGURE 5.2: Generation of the visual dictionary. Reprint from [238] ©2004
IEEE Computer Society Press.

5.5   α-Semantics Graph and Fuzzy Model for Repositories

Although we can take advantage of the semantics-oriented classification information from the training set, there are still issues not addressed yet. One is the semantic overlap between the classes. For example, one repository named "river" has affinities with the category named "lake". For certain users, the images in the repository "lake" are also interesting, although they pose a query image of "river". Another issue is the semantic uncertainty, which means that an image in one repository may also contain semantic objects inquired by the user although the repository is not for the semantics in which the user is interested. For instance, an image containing people in a "beach" repository is also relevant to users inquiring the retrieval of "people" images. To address these issues, we need to construct a model to explicitly describe the semantic relationships among images and the semantics representation for each repository.

5.5.1   α-Semantics Graph

The semantic relationships among images can be traced to a large extent in the feature space with statistical analysis. If the distribution of one semantic repository overlaps a great deal with another semantic repository in the feature space, it is a significant indication that these two semantic repositories have strong affinities. For example, "river" and "lake" have similar texture and shape attributes, e.g., "water" component. On the other hand, a repository having a loose distribution in the feature space has more uncertainty statistically compared with another repository having a more condensed distribution. In addition, the semantic similarity of two repositories can be measured by the shape of the feature distributions of the repositories as well as the distance between the corresponding distributions.

To describe these properties of semantic repositories quantitatively, we propose a metric to measure the scale, called *semantics correlation*, which reflects the relationship between two semantic repositories in the feature space. The semantics correlation is based on statistical measures of the shape of the repository distributions.

Perplexity. The perplexity of feature distributions of a repository reflects the uncertainty of the repository; it can be represented based on the entropy measurement [188]. Suppose there are k elements $s_1, s_2, \ldots, s_k$ in a set with probability distribution $P = \{p(s_1), p(s_2), \ldots, p(s_k)\}$. The entropy of the set is defined as

$$En(P) = -\sum_{i=1}^{k} p(s_i) \log p(s_i)$$

By Shannon's theorem [188], this is the lower bound on the average number of *bits per element (bpe)* required to encode a state of the set. For particular semantics represented in the images, it is difficult to precisely determine the probability of an image feature $p(s_i)$. Consequently, we use the statistics in the training semantic repository to estimate the probabilities. Since each image is represented as a 3-component vector $[C, T, S]$, the entropy of each repository, r_i, is defined as

$$H(r_i) = -\frac{1}{N_i} \sum_{j=1}^{N_j} P(C_j, T_j, S_j) \log P(C_j, T_j, S_j) \qquad (5.5)$$

where $P(C_i, T_i, S_i)$ is the joint occurrence probability of an image feature in the repository and N_i is the number of images in the repository. Assuming that color, texture, and shape properties are independent in the image representation, i.e., $P(C_j, T_j, S_j) = P(C_j)P(T_j)P(C_j)$ where $P(C_j)$, $P(T_j)$, and $P(S_j)$ are the occurrence probabilities of the single feature attribute in the repository, respectively, it follows that

$$H(r_i) = -\frac{1}{N_i} \sum_{j=1}^{N_i} P(C_j)P(T_j)P(S_j) \log\{P(C_j)P(T_j)P(S_j)\} \qquad (5.6)$$

As an analogy to the concept of *perplexity* [202, 177] for a text corpus, we define the *perplexity* of a semantic repository r_i in the image database as

$$\wp(r_i) = 2^{H(r_i)} \qquad (5.7)$$

which is an approximate measure of the homogeneity of the feature distributions in the repository r_i. The more perplex in the repository, the larger $\wp$, and vice versa.

Distortion. The distortion is a statistical measure to estimate the compactness degree of the repository. For each repository, r_i, distortion is defined as

$$D(r_i) = \frac{1}{N_i} \sqrt{\sum_{j=1}^{N_i} \|f_j - c_i\|^2} \qquad (5.8)$$

where f_j is the feature point j in this repository and c_i is the centroid of the repository. The distortion describes the distribution shape of repositories; i.e., the looser the repository, the larger D is defined.

Based on these statistical measures on the repositories, we propose a metric to describe the relationship between any two different repositories r_i and r_j, $i \neq j$, in the repository set Re. The metric, called *semantics correlation*, is a mapping $corr : Re \times Re \longrightarrow \mathbb{R}$. For any repository pair $\{r_i, r_j\}, i \neq j$, the semantics correlation is defined as

$$L_{i,j} = \frac{\sqrt{(D^2(r_i) + D^2(r_j))\wp(r_i)\wp(r_j)}}{\|c_i - c_j\|} \qquad (5.9)$$

$$corr_{i,j} = L_{i,j}/L_{max} \qquad (5.10)$$

where L_{max} is the maximal $L_{i,j}$ between any two different semantic reposito-ries, and $L_{max} = \max_{r_k, r_t \in Re, \, k \neq t}(L_{k,t})$. This definition of semantics correlation has the following properties:

- If the perplexity of a repository is large, which means that the homo-geneity degree of the repository is weak, it has a larger correlation with other repositories.

- If the distortion of a repository is large, which means that the repository is looser, it has a larger correlation with other repositories.

- If the inter-repository distance between two repositories is larger, the repository pair has a smaller correlation.

- The range of the semantics correlation is [0,1].

For convenience, the supplement of the semantics correlation for each semantic repository pair is defined as

$$disc_{i,j} = 1 - corr_{i,j} \qquad (5.11)$$

and is called the *semantics discrepancy* of the two different semantic reposito-ries. In this way, we give a quantitative measure of the relationship between any two different semantic repositories based on their distributions in the feature space.

With semantics correlation defined above, a graph is constructed on the repository space. We call the graph an α-semantics graph. It is defined as follows:

DEFINITION 5.1   *Given a semantic repository set $D = \{r_1, r_2, \ldots, r_m\}$, the semantics correlation function $corr_{i,j}$ defined on the set D, and a constant $\alpha \in \mathbb{R}$, a weighted undirected graph is called an α-semantics graph if it is constructed abiding by the following rules:*

- *The node set of the graph is the symbolic repository set.*

- *There is an edge between any nodes $i, j \in D$ if and only if $corr_{i,j} \geq \alpha$.*

- *The weight of the edge (i, j) is $corr_{i,j}$.*

The α-semantics graph uniquely describes the relationships between seman-tic repositories for an arbitrary α value. With a tuned α value, we can model a semantic repository based on its connected neighbors and corresponding edge weights in the α-semantics graph.

5.5.2 Fuzzy Model for Repositories

To address the semantic uncertainty and the semantic overlap problems, we propose a fuzzy model for each repository based on the constructed α-semantics graph. In this model, each semantics repository is defined as a fuzzy set while one particular image may belong to several semantic repositories.

A fuzzy set F on the feature space $\mathbb{R}^n$ is defined as a mapping $\mu_F : \mathbb{R}^n \rightarrow [0,1]$ named as the *membership function*. For any feature vector $f \in \mathbb{R}^n$, the value of $\mu_F(f)$ is called the degree of membership of f to the fuzzy set F (or, in short, the degree of membership to F). A value closer to 1 for $\mu_F(f)$ means the more representative the feature vector f is to the fuzzy set F (i.e., the semantic repository). For a fuzzy set F, there is a smooth transition for the degree of membership to F besides the hard cases $f \in F$ ($\mu_F(f) = 1$) and $f \notin F$ ($\mu_F(f) = 0$). It is clear that a fuzzy set degenerates to a conventional set if the range of μ_F is $\{0,1\}$ instead of $[0,1]$ (μ_F is then called the *characteristic function* of the set).

The most commonly used prototype membership functions are cone, trapezoidal, B-splines, exponential, Cauchy, and paired sigmoid functions [104]. Since we could not think of any intrinsic reason why one should be preferred to any other, we tested the cone, trapezoidal, exponential, and Cauchy functions on the system. In general, the performances of the exponential and the Cauchy functions are better than those of the cone and trapezoidal functions. Considering the computational complexity, we use the Cauchy functions because it requires much less computation. The Cauchy function is defined as

$$\mathcal{F}(x) = \frac{1}{1 + (\frac{\|x-v\|}{d})^{\beta}}$$

where d and $\beta \in \mathbb{R}$, $d > 0$, $\beta > 0$, v is the center location (point) of the fuzzy set, and d represents the width of the function and determines the shape (or smoothness) of the function. Collectively, d and β portray the grade of fuzziness of the corresponding fuzzy set. For fixed d, the grade of fuzziness increases as β decreases. If β is fixed, the grade of fuzziness increases with increased d. Figure 5.3 illustrates the Cauchy function in $\mathbb{R}$ with $v = 0$, $d = 36$, and β varying from 0.01 to 100. As we see, the Cauchy function approaches the characteristic function of an open interval (-36,36) when β goes to positive infinity. When β equals 0, the degree of membership for any element in $\mathbb{R}$ (except 0, whose degree of membership is always 1 in this example) is 0.5.

For each repository, the parameters v and d are determined based on the constructed α-semantics graph. The center point of each semantic repository r_i can be conveniently estimated by the mean vector, c_i, of the feature vectors in the repository. The width d_i is determined as follows:

$$d_i = \sum_{k=1}^{w} \|c_i - c_w\| corr_{i,w} \tag{5.12}$$

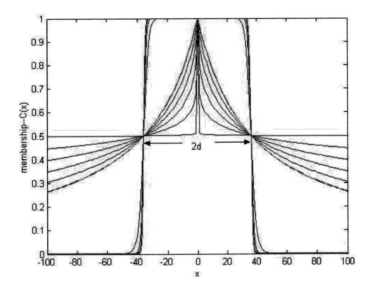

FIGURE 5.3: Cauchy function in one dimension.

where $\{c_1, c_2, \ldots, c_w\}$ is the set of the centroids of all connected nodes to the node r_i in the α-semantics graph and $\| \cdot \|$ is the Euclidean distance in $\mathbb{R}^n$. In other words, the width of the membership function for each repository is a semantics correlation weighted combination of the distance to its connected nodes in the α-semantics graph. Consequently, each repository r_i in the training set is modeled as a unique fuzzy set:

$$\mathcal{F}_i(f) = \frac{1}{1 + (\frac{\|f - c_i\|}{d_i})^\beta} \qquad (5.13)$$

Denoting the distance between a feature f and c_i as $dist$, the above equation can be equally presented as

$$\mathcal{F}_i(dist) = \frac{1}{1 + (\frac{dist}{d_i})^\beta} \qquad (5.14)$$

The experiments in Section 5.7 show that the performance changes insignificantly when β is in the interval $[0.7, 1.5]$ but degrades rapidly outside the interval. Thus, we set $\beta = 1$ in Equation 5.13 to simplify the computation.

5.6　Classification Based Retrieval Algorithm

With the three visual dictionaries ready, an order for the "keywords" in the visual dictionaries is determined and an index to each "keyword" is assigned. Given an image, for each feature attribute, replace it with the index of the "keyword" to which it is assigned in the corresponding visual dictionary. Hence, each image in the training set is represented as a tuple $Img[Color, Texture, Shape]$, while each attribute has a discrete value type in a limited domain.

To build a classification tree, the C4.5 algorithm [69] is applied on the training tuple sets transformed. We assume that each image in the training set belongs to only one semantic repository. The splitting attribute selection for each branch is based on the information gain ratio [69]. Associated with each leaf node of the classification tree is a ratio m/n, where m is the number of images classified to this node and n is the number of incorrectly classified images. This ratio is a measure of the classification accuracy of the classification tree for each class in the training image set.

Algorithm 8 lists the algorithm we have proposed for image mining and retrieval using an image query based on the classification with the fuzzy model for the repository to which the query image is classified.

In this algorithm, the repository is predicted by the classification tree for the query image. At the same time, a reference feature is determined by an inverse analysis from the classification accuracy; the reference feature's membership values of the semantic repositories of the neighborhood to the predicted repository in the α-semantics graph are determined. The intuition is illustrated in Figure 5.4. In this figure, two repositories modeled with the fuzzy set are shown. Every vector in the feature space is associated with the two repositories by the obtained membership values. These membership values are used as the sampling weights in the corresponding semantic repositories. In addition, since the algorithm is orthogonal to the distance metric $\mathcal{DM}$, different distance metrics $\mathcal{DM}$ can be used for different applications. In the evaluation experiments reported in Section 5.7, we use Euclidian distance as $\mathcal{DM}$ for its simplicity and effectiveness.

With this algorithm the images are mined and retrieved not only based on the repository the query image is classified to (which is called the *primary repository*) but also based on the semantics correlations between this primary repository and neighboring repositories in the α-semantics graph constructed. The percentage weight of images sampled in each potential relevant repository is determined by the corresponding classification accuracy and its fuzzy model. Intuitively, we give the majority of the share to the *primary repository*; the rest of the share to the connected repositories of the *primary repository* in the α-semantics graph is based on their semantics correlations with the *primary repository*. In other words, more weight of the share is given to the high

---

Algorithm 8 Image Querying Algorithm

Input: q, "keyword" tuple of the query image
Output: Images retrieved for the query image q
Method:

1: Initialization: returned image set $Result = \{\}$
2: $Q =$ the repository q is classified by the classification tree
3: $acc_Q =$ the accuracy of the classification associated with Q
4: $c_Q =$ the center of the repository Q
5: $d_Q =$ the width of the repository Q
6: Determine the distance $dist_Q$ between the reference feature rf and the
   center of the repository Q: $dist_Q = \sqrt[\beta]{(\frac{1}{acc_Q} - 1)d_Q}$
7: $SetS_Q =$ the images randomly sampled from the repository Q with percentage of acc_Q
8: $Result = Result \cup SetS_Q$
9: **for** Each node connected to the node Q in the α-semantics graph, V **do**
10:    **if** $\|c_V - c_Q\| >= dist_Q$ **then**
11:        $dist_V = \|c_V - c_Q\| - dist_Q$
12:    **else**
13:        $dist_V = dist_Q - \|c_V - c_Q\|$
14:    **end if**
15:    Determine the membership values of the rf, $\mathcal{F}_V(dist_V)$, using Equation 5.14
16:    The percentage sampling in the repository V, $PR_V = \mathcal{F}_V(rf)$
17:    $SetS_V =$ the images randomly sampled from the repository V with percentage of PR_V
18:    $Result = Result \cup SetS_V$
19: **end for**
20: Return the set $Result$ in a distance metric $\mathcal{DM}$ based rank with the query image

---

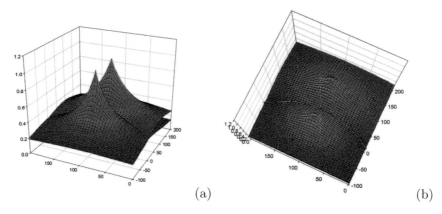

FIGURE 5.4: Illustration of two semantic repository models in the feature space. (a) Side view. (b) Top view; the dark curve represents part of the intersection curve. Reprint from [244] ©2004 ACM Press.

semantics-correlated repositories, while less weight of the share is given to the low semantics-correlated repositories. Consequently, we have solved for the semantic uncertainty and semantic overlap problems explicitly.

5.7   Experiment Results

We have implemented the methodology in a prototype system on a platform of a Pentium IV 2.0 GHz CPU with 256 MB memory. The image mining and retrieval evaluations were performed on a general-purpose color image database containing 10,000 images from a COREL collection of 96 semantic repositories. Each semantic repository had 85–120 images. Images in the same repository were often not all visually similar. Training images sampled from all 96 repositories were used to build the classification tree. Figure 5.5 shows a few samples of the images in the database.

5.7.1   Classification Performance on a Controlled Database

To provide quantitative evaluations on the performance of the image classification, we ran the prototype on a controlled subset of the COREL collection. This controlled database consisted of 10 image repositories (African people, beach, buildings, buses, dinosaurs, elephants, flowers, horses, mountains and glaciers, and food), each containing 100 pictures. Within this controlled database, we could assess classification performance reliably with the expected categorization accuracy because the repositories were semantically

FIGURE 5.5: Sample images in the database. The images in each column are assigned to one category. From left to right, the categories are Africa rural area, historical building, waterfalls, British royal event, and model portrait, respectively.

Table 5.1: Results of the classification tree based image classification experiments for the controlled database. Legend: A – Africa, B – Beach, C – Buildings, D – Buses, E – Dinosaurs, F – Elephants, G – Flowers, H – Horses, I – Mountains, and J – Foods. Reprint from [238] ©2004 IEEE Computer Society Press.

%	A	B	C	D	E	F	G	H	I	J
A	52	2	4	0	8	16	10	0	6	2
B	0	32	6	0	0	0	2	2	58	0
C	8	4	64	0	8	6	0	0	6	6
D	0	18	6	46	2	8	0	0	16	4
E	0	0	0	0	100	0	0	0	0	0
F	8	0	2	0	8	40	0	8	34	0
G	0	0	2	0	0	0	90	0	2	6
H	0	2	0	0	0	4	24	50	4	6
I	0	6	6	0	2	2	0	0	84	0
J	6	4	0	2	6	0	8	0	6	68

non-ambiguous and shared no semantic overlaps.

The classification performance of the constructed classification tree is compared with that of the nearest-neighbor classification method [36]. For both methods, 40 randomly chosen images for each repository are used to train the classifiers; the classification methods are then tested using the remaining 600 images outside the training set. The classification results of the proposed method and the raw feature based nearest-neighbor classification method [36] are shown in Tables 5.1 and 5.2, respectively. In both tables each row lists the percentage of the images in one repository classified to each of the 10 repositories. Numbers on the diagonal show the classification accuracy for every repository. The classification behavior of the proposed method is quite different from that of the nearest-neighbor method in the sense that the classification tree of the former is better than that of the latter because (i) the overall number of misclassifications between repositories is smaller, and (ii) the overall number of correct classifications is larger.

5.7.2 Classification Based Retrieval Results

For the 10,000 COREL image collection with 96 repositories, we have randomly shuffled the images in each repository and have taken 50% of them as the training set to train the image classifier. To evaluate the image mining and retrieval performance, 1,500 images are randomly selected from all repositories of the remaining 50% of the COREL collection as the query set.

Algorithm 8 is implemented as a prototype system, and Figure 5.6 shows the interface of the prototype system. We have invited a group of 5 users to participate in the evaluations. The participants consist of Computer Science

Table 5.2: Results of the nearest-neighbor based image classification experiments for the controlled database. Legend: A – Africa, B – Beach, C – Buildings, D – Buses, E – Dinosaurs, F – Elephants, G – Flowers, H – Horses, I – Mountains, and J – Foods. Reprint from [238] ©2004 IEEE Computer Society Press.

%	A	B	C	D	E	F	G	H	I	J
A	**33**	11	10	0	7	12	6	8	10	3
B	3	**35**	4	0	0	20	1	13	14	10
C	7	7	**45**	3	5	17	0	3	13	0
D	4	13	7	**40**	0	8	2	4	18	4
E	0	0	1	0	**88**	0	6	5	0	0
F	3	0	6	0	2	**46**	0	9	27	7
G	1	1	2	8	0	0	**78**	0	2	10
H	1	3	0	7	0	11	18	**34**	15	11
I	4	7	9	0	2	4	0	0	**69**	5
J	10	4	5	6	3	6	10	0	23	**33**

graduate students as well as lay people outside the Computer Science Department. The relevancy of the retrieved images is subjectively examined by the users, and the retrieval accuracy includes the average values across all query sessions.

Before we evaluate the prototype system, an appropriate α has to be decided. For the extreme case $\alpha = 0$, each node is connected to all other nodes in the 0-semantics graph (all the repositories are treated as semantics-related to each other); for $\alpha = 1$, each node is isolated (with no edges connected to other nodes), and the 1-semantics graph is degraded to a repository set. In the experiments we compute pair-wise semantics correlations $corr_{i,j}$ for all the repository pairs in the training set; the third quartile, which is obtained as 0.649 for the training set, is used as the α in the prototype.

Figure 5.7 shows an excerpted α-semantics graph example with $\alpha = 0.649$ for the repositories in the training set. The annotation of each repository is labeled on its node. The length of each edge between two nodes in the figure is proportional to the *semantics discrepancy* between the two connecting repositories. It is noticeable that the semantic uncertainty and the semantic overlap among repositories described in Section 5.5.1 are measured explicitly. For example, for the "outdoor scene" repository, repository "castle" is more semantics-correlated than repository "beach"; repository "waterfall" has strong semantics correlations with repositories "fishing", "rafting", and "beach"; repository "peasant life" is connected to repositories "outdoor scene" and "fashion model". These semantics correlations measured in the feature space among repositories agree well with the subjective perceptions of the image contents.

Figure 5.8 shows three test images with 1, 3, and 7 repositories connected in

FIGURE 5.6: Interface of the prototype system.

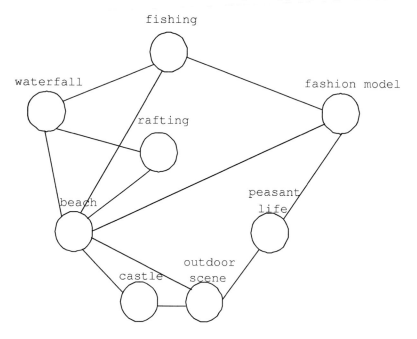

FIGURE 5.7: An example of an α-semantics graph with $\alpha = 0.649$. Reprint from [244] ©2004 ACM Press.

(a)   (b)   (c)

FIGURE 5.8: Three test images. (a) This image is associated with a single repository in an α-semantics graph. (b) This image is associated with 3 repositories. (c) This image is associated with 7 repositories. Reprint from [244] ©2004 ACM Press.

the constructed α-semantics graph, respectively. The *primary repository* assigned to Figure 5.8(a) is the repository "china", which is correct, without any edges connected. Figure 5.8(b) is assigned the *primary repository* "people", and two repositories, "building" and "outdoor scene", are connected with the *primary repository* with the corresponding semantics correlations 0.652 and 0.723, respectively. Based on the subjective observation, "building" is not relevant, while the *primary repository* "people" and the other connected repository, "outdoor scene", are. The *primary repository* of Figure 5.8(c) is "winter season" with connections to repositories "building", "beach", "European town", "mountain", "sea shore", and "vacation resort" in the α-semantics graph. Although the *primary repository* "winter season" assigned to this image by the classification tree is not semantically relevant, there are 4 semantically relevant repositories ("building", "European town", "sea shore", and "vacation resort") connected with the *primary repository* ("winter season"). Thus, the retrieval accuracy is significantly improved by incorporating these repositories into the fuzzy model described in Section 5.5.

To evaluate the effectiveness of the semantics correlation measurement and the fuzzy model for repositories, we have compared the retrieval precision with and without the α-semantics graph. Figure 5.9 shows the results. From the figure, it is evident that the α-semantics graph and the derived fuzzy model for repositories improve the precision significantly. These results substantiate the motivations: by explicitly addressing the semantic uncertainty and the semantic overlap, the classification errors can be substantially reduced for image mining and retrieval.

To evaluate the influence of the error rate of the classification tree, we have documented the statistics of the classifications and the corresponding retrieval precisions for the testing and training sets. Three evaluation statistics are documented. They are:

Average classification error rate: The average rate at which a query image is misclassified.

Average classification accuracy: The average value of the classification

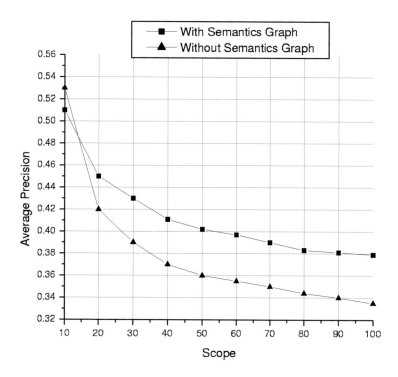

FIGURE 5.9: Average precision comparison with/without the α-semantics graph. Reprint from [244] ©2004 ACM Press and from [238] ©2004 IEEE Computer Society Press.

Table 5.3: The classification statistics of our method and the nearest-neighbor method.

	Average classification error rate	Average classification accuracy
The classification tree	0.235	0.868
Nearest-neighbor method	0.307	0.775

Table 5.4: The classification and retrieval precision statistics. Reprint from [244] ©2004 ACM Press and from [238] ©2004 IEEE Computer Society Press.

	Average classification accuracy	Average retrieval precision
Correctly classified images	0.902	0.672
Incorrectly classified images	0.815	0.343

accuracy for training images in all repositories.

Average retrieval precision: The average ratio of relevant images in the top 50 retrieved images for every query image.

The results are shown in Table 5.3 and Table 5.4. We have also compared the classification performances of the classification tree and the nearest-neighbor classification method on the testing image database, as shown in Table 5.3. In the comparison, the classification tree is consistently better than the nearest-neighbor classification in that: (i) the average classification error for testing images is smaller; and (ii) the average classification accuracy for training images is larger.

For an image retrieval example, Figure 5.10 shows the top 16 retrieved images by the prototype system for an query image from the repository "city skyline". The retrieval precision is satisfactory; 15 of 16 of the top returned images are relevant.

Considering that it is difficult to design a fair comparison with very few existing classification-based image mining and retrieval methods, we have compared the effectiveness of the proposed method with that of UFM [47]. UFM is a methodology based on fuzzified region representation to build region-to-region similarity measures for image retrieval. It does not address semantics of images explicitly. We compare the proposed approach with UFM because UFM is available to us and also because UFM reflects the performance of the state-of-the-art image mining and retrieval. The results are shown in

FIGURE 5.10: Query result for an image from the repository "city skyline".
15 out of the top 16 returned images are relevant.

Figure 5.11. From the comparisons, it is clear that the proposed method is
superior to UFM in both absolute precision and potential (attenuation trend).

Another advantage of the proposed method is its high online query efficiency. In most state-of-the-art image mining and retrieval systems in the
research literature, the search is performed linearly. In other words, the computation complexity is $O(n)$ for a image database with n images. In the
proposed method, the average computation complexity is $O(\log m)$ for image classification and $O(w)$ for image similarity calculation, where m is the
number of image repositories and w is the average number of images in a
repository. Since $w = \frac{n}{m}$, the overall complexity is $O(\log m + \frac{n}{m})$. In general,
$m << n$; hence, with image classification the computation complexity of this
method is much more tractable than that of the linear search methods. This
conclusion is also observed in the experiments. The average query time for
returning the top 30 images is less than 0.5 second in the reported platform.

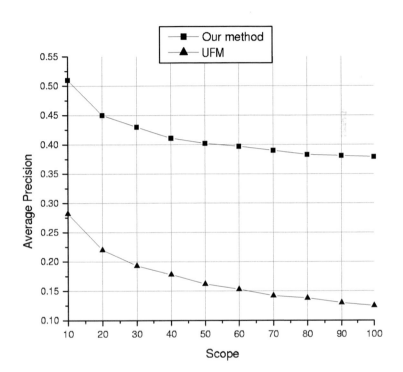

FIGURE 5.11: Average precision comparison between the proposed method and UFM. Reprint from [238] ©2004 IEEE Computer Society Press.

5.8   Summary

In this chapter, we have presented an image classification based approach in the traditional architecture of the content based mining and retrieval in a large image database. A semantics correlation based structure called the α-semantics graph is proposed to represent the semantic uncertainty and the semantic overlap explicitly. Founded on the α-semantics graph, each semantic repository is modeled as a fuzzy set which captures the statistical distribution in the feature space. With the generation of a multiple-features (color, texture, and shape) supported visual dictionary, a classification tree is trained using a provided training set. A unique image mining and retrieval algorithm is developed through integrating the classification results and the fuzzy model for each repository. With the effective supervised learning applied to the image database and the precise modeling of image semantic repositories, the proposed methodology inaugurates a new generation of content-based image mining and retrieval approaches, that aims at achieving a more semantics-relevant performance.

Chapter 6

Image Database Modeling – Latent Semantic Concept Discovery

6.1 Introduction

This chapter addresses image database modeling in general and, in particular, focuses on developing a hidden semantic concept discovery methodology to address effective semantics-intensive image data mining and retrieval. In the approach proposed in this chapter, each image in the database is segmented into regions associated with homogenous color, texture, and shape features. By exploiting regional statistical information in each image and employing a vector quantization method, a uniform and sparse region-based representation is achieved. With this representation a probabilistic model based on the statistical-hidden-class assumptions of the image database is obtained, to which the Expectation-Maximization (EM) technique is applied to discover and analyze semantic concepts hidden in the database. An elaborated mining and retrieval algorithm is designed to support the probabilistic model. The semantic similarity is measured through integrating the posterior probabilities of the transformed query image, as well as a constructed negative example, to the discovered semantic concepts. The proposed approach has a solid statistical foundation; the experimental evaluations on a database of 10,000 general-purpose images demonstrate the promise and the effectiveness of the proposed approach.

The rest of this chapter is organized as follows. Section 6.2 gives background information regarding why it is necessary to propose and develop the latent semantic concept discovery approach to model an image database and reviews the related work in the literature. Section 6.3 introduces the region feature extraction method and the region based image representation scheme used in developing this latent semantic concept discovery approach. Section 6.4 then presents the proposed probabilistic region–image–concept model and the hidden semantic concept discovery procedure using the Expectation-Maximization method developed in this approach. Section 6.5 presents the posterior probability based image similarity measure scheme and the supportive relevance feedback based mining and retrieval algorithm. An analysis of the characteristics of the proposed approach and its uniqueness in comparison with the existing region based image data mining and retrieval methods is provided in Section 6.6. Section 6.7 reports the experimental evaluations of this proposed approach in comparison with a state-of-the-art method from the literature and demonstrates the superior performance of this approach in image data mining and retrieval. Finally, this chapter is concluded in Section 6.8.

6.2 Background and Related Work

As stated before, large collections of images have become available to the public, from photo collections to Web pages or even video databases. To effectively mine or retrieve such a large collection of imagery data is a huge challenge. After more than a decade of research, it has been found that content based image data mining and retrieval are a practical and satisfactory solution to this challenge. At the same time, it is also well known that the performance of the existing approaches in the literature is mainly limited by the *semantic gap* between low-level features and high-level semantic concepts [192]. In order to reduce this gap, region based features (describing object level features), instead of raw features of the whole image, to represent the visual content of an image are widely used [36, 212, 119, 47].

In contrast to traditional approaches [112, 80, 166], which compute global features of images, the region based methods extract features of the segmented regions and perform similarity comparisons at the granularity of regions. The main objective of using region features is to enhance the ability to capture and represent the focus of users' perception of the image content.

One important issue significantly affecting the success of an image data mining methodology is how to compare two images, i.e., the definition of the image similarity measurement. A straightforward solution adopted by most early systems [36, 142, 221] is to use individual region-to-region similarity as the basis of the comparisons. When using such schemes, the users are forced to select a limited number of regions from a query image in order to start a query session. As discussed in [212], due to the uncontrolled nature of the visual content in an image, automatically and precisely extracting image objects is still beyond the reach of the state-of-the-art in computer vision. Therefore, these systems tend to partition one object into several regions, with none of them being representative for the object. Consequently, it is often difficult for users to determine which regions should be used for their interest.

To provide users a simpler querying interface and to reduce the influence of inaccurate segmentation, several image-to-image similarity measurements that combine information from all of the regions have been proposed [91, 212, 47]. Such systems only require users to impose a query image and therefore relieve the users from making the puzzling decisions. For example, the SIMPLIcity system [212] uses integrated region matching as its image similarity measure. By allowing a many-to-many relationship of the regions, the approach is robust to inaccurate segmentation. Greenspan et al [92] propose a continuous probabilistic framework for image matching. In this framework, each image is represented as a Gaussian mixture distribution, and images are compared and matched via a probabilistic measure of similarity between distributions. Improved image matching results are reported.

Ideally, what we strive to measure is the *semantic similarity*, which physically is very difficult to define, or even to describe. The majority of the existing methodologies do not explicitly connect the extracted features with the pursued semantics reflected in the visual content. They define region-to-region and/or image-to-image similarities to attempt to approximate the semantic similarity. However, the approximation is typically heuristic and consequently not reliable and effective. Thus, the retrieval and mining accuracies are rather limited.

To deal with the inaccurate approximation problem, several research efforts have been attempted to link regions to semantic concepts by supervised learning. Barnard et al proposed several statistical models [14, 70, 15] which connect *image blobs* and linguistic words. The objective is to predict words associated with whole images (auto-annotation) and corresponding to particular image regions (region naming). In their approaches, a number of models are developed for the joint distribution of image regions and words. The models are multi-modal and correspondence extensions to Hofmann's hierarchical clustering aspect model [102, 103, 101], a translation model adapted from statistical machine translation, and a multi-modal extension to the mixture of latent Dirichlet allocation models [22]. The models are used to automatically annotate testing images, and the reported performance is promising. Recognizing that these models fail to exploit spatial context in the images and words, Carbonetto et al augmented the models such that spatial relationships between regions are learned. The model proposed is more expressive in the sense that the spatial correspondences are incorporated into the joint probability learning [34, 35], which improves the accuracy of object recognition in image annotation. Recently, Feng et al proposed a Multiple Bernoulli Relevance Model (MBRM) [75] for image-word association, which is based on the Continuous-space Relevance Model (CRM) proposed by [117]. In the MBRM model, the word probabilities are estimated using a multiple Bernoulli model and the image feature probabilities using a non-parametric kernel density estimate.

We argue that for all the feature based image mining and retrieval methods, the semantic concepts related to the content of the images are always hidden. By hidden, we mean (1) objectively, there is no direct mapping from the numerical image features to the semantic meanings in the images, and (2) subjectively, given the same region, there are different corresponding semantic concepts, depending on different context and/or different user interpretations.

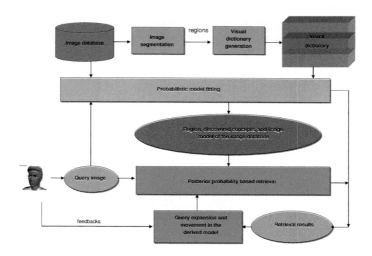

FIGURE 6.1: The architecture of the latent semantic concept discovery based image data mining and retrieval approach. Reprint from [243] ©2007 IEEE Signal Processing Society Press.

This observation justifies the need to discover the hidden semantic concepts that is a key step toward effective image retrieval.

In this chapter, we propose a probabilistic approach to addressing the hidden semantic concept discovery. A region-based sparse but uniform image representation scheme is developed (unlike the block-based uniform representation in [255], region-based representation is more effective for image mining and retrieval due to the fact that humans pay more attention to objects than blocks in an image), which facilitates the indexing scheme based on a region-image-concept probabilistic model with validated assumptions. This model has a solid statistical foundation and is intended for the objective of semantics-intensive image retrieval. To describe the semantic concepts hidden in the region and image distributions of a database, the Expectation-Maximization (EM) technique is used. With a derived iterative procedure, the posterior probabilities of each region in an image for the hidden semantic concepts are quantitatively obtained, which act as the basis for the *semantic similarity* measure for image mining and retrieval. Therefore, the effectiveness is improved as the similarity measure is based on the discovered semantic concepts, which are more reliable than the region features used in most of the existing systems in the literature. Figure 6.1 shows the architecture of the proposed approach. This work is an extension of the previous work [240].

Different from the models reviewed above, the model and the approach we propose and present here do not require training data; we formulate a generative model to discover the clusterings in a probabilistic scheme by unsupervised learning. In this model, the regions and images are connected through a hidden layer — the concept layer, which constitutes the basis of the image similarity measures. In addition, users' relevance feedback is incorporated into the model fitting procedure such that the subjectivity in image mining and retrieval is addressed explicitly and the model fitting is customized toward users' querying needs.

6.3   Region Based Image Representation

In the proposed approach, the query image and images in a database are first segmented into homogeneous color-texture regions. Then representative properties are extracted for every region by incorporating multiple features, specifically, color, texture, and shape properties. Based on the extracted regions, a visual token catalog is generated to explore and exploit the content similarities of the regions, which facilitates the indexing and mining scheme based on the region-image-concept probabilistic model elaborated in Section 6.4.

6.3.1   Image Segmentation

To segment an image, the system first partitions the image into blocks of 4 by 4 pixels to compromise between the texture effectiveness and the computation time. Then a feature vector consisting of nine features from each block is extracted. Three of the features are average color components in the 4 by 4 pixel size block; we use the LAB color space due to its desired property that the perceptual color difference is proportional to the numerical difference. The other six features are the texture features extracted using wavelet analysis.

To extract texture information of each block, we apply a set of Gabor filters [145], which are shown to be effective for image indexing and retrieval [143], to the block to measure the response. The Gabor filters measure the two-dimensional wavelets. The discretization of a two-dimensional wavelet applied to the blocks is given by

$$W_{mlpq} = \int \int I(x,y)\psi_{ml}(x - p\triangle x, y - q\triangle y)dxdy \qquad (6.1)$$

where I denotes the processed block; $\triangle x$ and $\triangle y$ denote the spatial sampling rectangle; p, q are image positions; and m, l specify the scale and orientation of the wavelets. The base function $\psi_{ml}(x, y)$ is given by

$$\psi_{ml}(x, y) = a^{-m}\psi(\widetilde{x}, \widetilde{y}) \tag{6.2}$$

where

$$\widetilde{x} = a^{-m}(x \cos \theta + y \sin \theta)$$

$$\widetilde{y} = a^{-m}(-x \sin \theta + y \cos \theta)$$

denote a dilation of the mother wavelet (x, y) by a^{-m}, where a is the scale parameter, and a rotation by $\theta = l \times \triangle\theta$, where $\triangle\theta = 2\pi/V$ is the orientation sampling period; V is the number of orientation sampling intervals.

In the frequency domain, with the following Gabor function as the mother wavelet, we use this family of wavelets as our filter bank:

$$\begin{aligned}\Psi(u, v) &= \exp\left\{-2\pi^2(\sigma_x^2 u^2 + \sigma_y^2 v^2)\right\} \otimes \delta(u - W) \\ &= \exp\left\{-2\pi^2(\sigma_x^2(u - W)^2 + \sigma_y^2 v^2)\right\} \\ &= \exp\left\{-\frac{1}{2}(\frac{(u - W)^2}{\sigma_u^2} + \frac{v^2}{\sigma_v^2})\right\}\end{aligned} \tag{6.3}$$

where $\otimes$ is a convolution symbol, $\delta(.)$ is the impulse function, $\sigma_u = (2\pi\sigma_x)^{-1}$, and $\sigma_v = (2\pi\sigma_y)^{-1}$; σ_x and σ_y are the standard deviations of the filter along the x and y directions, respectively. The constant W determines the frequency bandwidth of the filters.

Applying the Gabor filter bank to the blocks, for every image pixel (p, q), in U (the number of scales in the filter bank) by V array of responses to the filter bank, we only need to retain the magnitudes of the responses:

$$F_{mlpq} = |W_{mlpq}| \quad m = 0, \ldots, U - 1, \ l = 0, \ldots V - 1 \tag{6.4}$$

Hence, a texture feature is represented by a vector, with each element of the vector corresponding to the energy in a specified scale and orientation sub-band w.r.t. a Gabor filter. In the implementation, a Gabor filter bank of 3 orientations and 2 scales is used for each image in the database, resulting in a 6-dimensional feature vector (i.e., 6 means for $|W_{ml}|$) for the texture representation.

After we obtain feature vectors for all blocks, we perform normalization on both color and texture features such that the effects of different feature ranges are eliminated. Then a k-means based segmentation algorithm, similar to that used in [47], is applied to clustering the feature vectors into several classes, with each class corresponding to one region in the segmented image.

Figure 6.2 gives four examples of the segmentation results of images in the database, which show the effectiveness of the segmentation algorithm employed.

After the segmentation, the edge map is used with the water-filling algorithm [253] to describe the shape feature for each region due to its reported effectiveness and efficiency for image mining and retrieval [154]. A

FIGURE 6.2: The segmentation results. Left column shows the original images; right column shows the corresponding segmented images with the region boundary highlighted.

6-dimensional shape feature vector is obtained for each region by incorporating the statistics defined in [253], such as the filling time histogram and the fork count histogram. The mean of the color-texture features of all the blocks in each region is determined to combine with the corresponding shape feature as the extracted feature vector of the region.

6.3.2   Visual Token Catalog

Since the region features $f \in \mathbb{R}^n$, it is necessary to perform regularization on the region property set such that they can be indexed and mined efficiently. Considering that many regions from different images are very similar in terms of the features, vector quantization (VQ) techniques are required to group similar regions together. In the proposed approach, we create a visual token catalog for region properties to represent the visual content of the regions. There are three advantages to creating such a visual token catalog. First, it improves mining and retrieval robustness by tolerating minor variations among visual properties. Without the visual token catalog, since very few feature values are exactly shared by different regions, we would have to consider feature vectors of all the regions in the database. This makes it not effective to compare the similarity among regions. However, based on the visual token catalog created, low-level features of regions are quantized such that images can be represented in a way resistant to perception uncertainties [47]. Second, the region-comparison efficiency is significantly improved by mapping the expensive numerical computation of the distances between region features to the inexpensive symbolic computation of the differences between "code words" in the visual token catalog. Third, the utilization of the visual token catalog reduces the storage space without sacrificing the accuracy.

We create the visual token catalog for region properties by applying the Self-Organization Map (SOM) [130] learning strategy. SOM is ideal for this problem, as it projects the high-dimensional feature vectors to a 2-dimensional plane through mapping similar features together while separating different features at the same time. The SOM learning algorithm we have used is competitive and unsupervised. The nodes in a 2-dimensional array become specifically tuned to various classes of input feature patterns in an orderly fashion.

A procedure is designed to create "code words" in the dictionary. Each "code word" represents a set of visually similar regions. The procedure follows 4 steps:

1. Performing the Batch SOM learning [130] algorithm on the region feature set to obtain the visualized model (node status) displayed on a 2-dimensional plane map. The distance metric used is Euclidean for its simplicity.

2. Regarding each node as a "pixel" in the 2-dimensional plane map such that the map becomes a binary lattice with the value of each pixel i

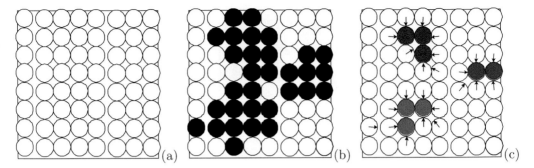

FIGURE 6.3: Illustration of the procedure: (a) the initial map; (b) the binary lattice obtained after the SOM learning is converged; (c) the labeled object on the final lattice. The arrows indicate the objects that the corresponding nodes belong to. Reprint from [243] ©2007 IEEE Signal Processing Society Press.

defined as follows:

$$p(i) = \begin{cases} 0 \text{ if } count(i) \geq t \\ 1 \text{ else} \end{cases}$$

where $count(i)$ is the number of features mapped to node i and the constant t is a preset threshold. Pixel value 0 denotes the objects, while pixel value 1 denotes the background.

3. Performing the morphological erosion operation [38] on the resulting lattice to make sparse connected objects in the image disjointed. The size of the erosion mask is determined to be the minimum to make two sparsely connected objects separated.

4. With connected component labeling [38], we assign each separated object a unique ID, a "code word". For each "code word", the mean of all the features associated with it is determined and stored. All "code words" constitute the visual token catalog to be used to represent the visual properties of the regions.

Figure 6.3 illustrates this procedure on a portion of the map we have obtained.

Simple yet effective Euclidean distance is used in the SOM learning to determine the "code word" to which each region belongs. The proof of the convergence of the SOM learning process in the 2-dimensional plane map is given in [129]. The details about the selection of the parameters are also covered in [129]. Each labeled component represents a region feature set among which the intra-distance is low. The extent of similarity in each "code word" is controlled by the parameters in the SOM algorithm and the threshold t. With this procedure, the number of the "code words" is adaptively determined and the similarity-based feature grouping is achieved. The experiments reported

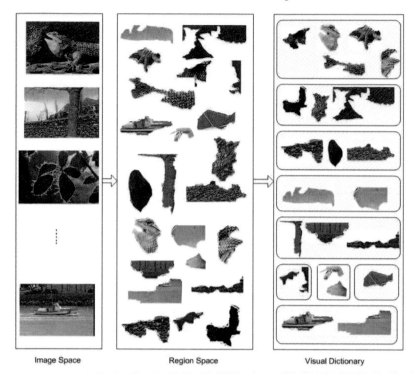

FIGURE 6.4: The process of the generation of the visual token catalog. Reprint from [243] ©2007 IEEE Signal Processing Society Press and from [240] ©2004 IEEE Computer Society Press.

in Section 6.7 show that the visual token catalog created captures the clustering characteristics existing in the feature set well. We note that the threshold t is highly correlated to the number of the "code words" generated; it is determined empirically by balancing the efficiency and the accuracy. We discuss the issue of choosing the appropriate number of the "code words" in the visual token catalog in Section 6.7. Figure 6.4 shows the process of the generation of the visual token catalog. Each rounded rectangle in the third column of the figure is one "code word" in the dictionary.

For each region of an image in the database, the "code word" that it is associated with is identified and the corresponding index in the visual token catalog is stored, while the original feature of this region is discarded. For the region of a new image, the closest entry in the dictionary is found and the corresponding index is used to replace its feature. In the rest of this chapter, we use the terminologies *region* and *"code word"* interchangeably; they both denote an entry in the visual token catalog equivalently.

Based on the visual token catalog, each image is represented in a uniform vector model. In this representation, an image is a vector with each dimension corresponding to a "code word". More formally, the uniform representation $\vec{I}_u$ of an image I is a vector $\vec{I}_u = \{w_1, w_2, \ldots, w_M\}$, where M is the number of the "code words" in the visual token catalog. For a "code word" $C_i, 1 \leq i \leq M$, if there exists a region R_j of I that corresponds to it, then $w_i = W_{Rj}$ for $\vec{I}_u$, where W_{Rj} is the number of the occurrences of R_j in the image I; otherwise, $w_i = 0$. This uniform representation is sparse, for an image usually contains a few regions compared with the number of the "code words" in the visual token catalog. Based on this representation of all the images, the database is modeled as a $M \times N$ "code word"-image matrix which records the occurrences of every "code word" in each image, where N is the number of the images in the database.

6.4   Probabilistic Hidden Semantic Model

To achieve the automatic semantic concept discovery, a region-based probabilistic model is constructed for the image database with the representation of the "code word"-image matrix. The probabilistic model is analyzed by the Expectation-Maximization (EM) technique [58] to discover the latent semantic concepts, which act as a basis for effective image mining and retrieval via the concept similarities among images.

6.4.1   Probabilistic Database Model

With a uniform "code word" vector representation for each image in the database, we propose a probabilistic model. In this model, we assume that the specific (region, image) pairs are known i.i.d. samples from an unknown distribution. We also assume that these samples are associated with an unobserved *semantic concept* variable $z \in Z = \{z_1, \ldots, z_K\}$, where K is the number of concepts to be discovered. Each observation of one region ("code word") $r \in R = \{r_1, \ldots, r_M\}$ in an image $g \in G = \{g_1, \ldots, g_N\}$ belongs to one concept class z_k. To simplify the model, we have two further assumptions. First, the observation pairs (r_i, g_j) are generated independently. Second, the pairs of random variables (r_i, g_j) are conditionally independent given the respective hidden concept z_k, i.e., $P(r_i, g_j | z_k) = P(r_i | z_k) P(g_j | z_k)$. Intuitively, these two assumptions are reasonable, which are further validated by the experimental evaluations. The region and image distribution may be treated as a randomized data generation process, described as follows:

- Choose a concept with probability $P(z_k)$;

- Select a region $r_i \in R$ with probability $P(r_i | z_k)$; and

- Select an image $g_j \in G$ with probability $P(g_j | z_k)$.

As a result, one obtains an observed pair (r_i, g_j), while the concept variable z_k is discarded.

Based on the theory of the generative model [150] (see Chapter 3), the above process is equivalent to the following:

- Select an image g_j with probability $P(g_j)$;

- Select a concept z_k with probability $P(z_k|g_j)$;

- Generate a region r_i with probability $P(r_i|z_k)$.

Translating this process into a joint probability model results in the expression

$$P(r_i, g_j) = P(g_j)P(r_i|g_j)$$

$$= P(g_j) \sum_{k=1}^{K} P(r_i|z_k)P(z_k|g_j) \qquad (6.5)$$

Inverting the conditional probability $P(z_k|g_j)$ in Equation 6.5 with the application of Bayes' rule results in

$$P(r_i, g_j) = \sum_{k=1}^{K} P(z_k)P(r_i|z_k)P(g_j|z_k) \qquad (6.6)$$

Following the likelihood principle, one determines $P(z_k)$, $P(r_i|z_k)$, and $P(g_j|z_k)$ by the maximization of the log-likelihood function

$$\mathcal{L} = \log P(R, G) = \sum_{i=1}^{M} \sum_{j=1}^{N} n(r_i, g_j) \log P(r_i, g_j) \qquad (6.7)$$

where $n(r_i, g_j)$ denotes the number of the regions r_i that occurred in image g_j. From Equations 6.7 and 6.5 we derive that the model is a statistical mixture model [150], which can be resolved by applying the EM technique [58].

6.4.2 Model Fitting with EM

One powerful procedure for maximum likelihood estimation in hidden variable models is the EM method [58]. EM alternates in two steps iteratively: (i) an expectation (E) step where posterior probabilities are computed for the hidden variable z_k, based on the current estimates of the parameters, and (ii) a maximization (M) step, where parameters are updated to maximize the expectation of the complete-data likelihood $\log P(R, G, Z)$ for the given posterior probabilities computed in the previous E-step.

Applying Bayes' rule with Equation 6.5, we determine the posterior probability for z_k under (r_i, g_j):

$$P(z_k|r_i, g_j) = \frac{P(z_k)P(g_j|z_k)P(r_i|z_k)}{\sum_{k'=1}^{K} P(z_{k'})P(g_j|z_{k'})P(r_i|z_{k'})} \qquad (6.8)$$

The expectation of the complete-data likelihood $\log P(R, G, Z)$ for the estimated $P(Z|R, G)$ derived from Equation 6.8 is

$$E\{\log P(R, G, Z)\} = \sum_{(i,j)=1}^{K} \sum_{i=1}^{M} \sum_{j=1}^{N} n(r_i, g_j) \log \left[P(z_{i,j}) P(g_j|z_{i,j}) P(r_i|z_{i,j}) \right] P(Z|R, G)$$

$$(6.9)$$

where

$$P(Z|R, G) = \prod_{m=1}^{M} \prod_{n=1}^{N} P(z_{m,n}|r_m, g_n)$$

In Equation 6.9 the notation $z_{i,j}$ is the concept variable that is associated with the region-image pair (r_i, g_j). In other words, (r_i, g_j) belongs to concept z_t where $t = (i, j)$.

With the normalization constraint $\sum_{(i,j)=1}^{K} P(z_{i,j}|r_i, g_j) = 1$, Equation 6.9 further becomes:

$$E\{\log P(R, G, Z)\} = \sum_{l=1}^{K} \sum_{i=1}^{M} \sum_{j=1}^{N} n(r_i, g_j) \log[P(r_i|z_l) P(g_j|z_l)] P(z_l|r_i, g_j) +$$

$$+ \sum_{l=1}^{K} \sum_{i=1}^{M} \sum_{j=1}^{N} n(r_i, g_j) \log[P(z_l)] P(z_l|r_i, g_j) \qquad (6.10)$$

Maximizing Equation 6.10 with Lagrange multipliers to $P(z_l)$, $P(r_u|z_l)$, and $P(g_v|z_l)$, respectively, under the following normalization constraints

$$\sum_{k=1}^{K} P(z_k) = 1 \qquad (6.11)$$

$$\sum_{k=1}^{K} P(z_k|r_i, g_j) = 1 \qquad (6.12)$$

$$\sum_{i=1}^{M} P(r_i|z_l) = 1 \qquad (6.13)$$

for any r_i, g_j, and z_l, the parameters are determined as

$$P(z_k) = \frac{\sum_{i=1}^{M} \sum_{j=1}^{N} n(r_i, g_j) P(z_k|r_i, g_j)}{\sum_{i=1}^{M} \sum_{j=1}^{N} u(r_i, g_j)} \qquad (6.14)$$

$$P(r_u|z_l) = \frac{\sum_{j=1}^{N} n(r_u, g_j) P(z_l|r_u, g_j)}{\sum_{i=1}^{M} \sum_{j=1}^{N} u(r_i, g_j) P(z_l|r_i, g_j)} \qquad (6.15)$$

$$P(g_v|z_l) = \frac{\sum_{i=1}^{M} n(r_i, g_v) P(z_l|r_i, g_v)}{\sum_{i=1}^{M} \sum_{j=1}^{N} u(r_i, g_j) P(z_l|r_i, g_j)} \qquad (6.16)$$

Alternating Equation 6.8 with Equations 6.14–6.16 defines a convergent procedure that approaches a local maximum of the expectation in Equation 6.10. The initial values for $P(z_k)$, $P(g_j|z_k)$, and $P(r_i|z_k)$ are set to be the same as if the distributions of $P(Z)$, $P(G|Z)$, and $P(R|Z)$ are the uniform distributions; in other words, $P(z_k) = 1/K$, $P(r_i|z_k) = 1/M$, and $P(g_j|z_k) = 1/N$. We have found in the experiments that different initial values only affect the number of iterative steps to the convergence but have no effects on the converged values of them.

6.4.3   Estimating the Number of Concepts

The number of concepts, K, must be determined in advance to initiate the EM model fitting. Ideally, we would like to select the value of K that best represents the number of the semantic classes in the database. One readily available notion of the goodness of the fitting is the log-likelihood. Given this indicator, we apply the Minimum Description Length (MDL) principle [174, 175] to select the best value of K. This can be operationalized as follows [175]: choose K to maximize

$$\log(P(R, G)) - \frac{m_K}{2} \log(MN) \qquad (6.17)$$

where the first term is expressed in Equation 6.7 and m_K is the number of the free parameters needed for a model with K mixture components. In the case of the proposed probabilistic model, we have

$$m_K = (K - 1) + K(M - 1) + K(N - 1) = K(M + N - 1) - 1$$

As a consequence of this principle, when models using two values of K fit the data equally well, the simpler model is selected. In the database used in the experiments reported in Section 6.7, K is determined through maximizing Equation 6.17.

6.5   Posterior Probability Based Image Mining and Retrieval

Based on the probabilistic model, we can derive the posterior probability of each image in the database for every discovered concept by applying Bayes' rule as

$$P(z_k|g_j) = \frac{P(g_j|z_k)P(z_k)}{P(g_j)} \qquad (6.18)$$

which can be determined using the estimations in Equations 6.14–6.16. The posterior probability vector $P(Z|g_j) = [P(z_1|g_j), P(z_2|g_j), \ldots, P(z_K|g_j)]^T$ is used to quantitatively describe the semantic concepts associated with the image g_j. This vector can be treated as a representation of g_j (which originally has a representation in the M-dimensional "code word" space) in the K-dimensional *concept space* determined using the estimated $P(z_k|r_i, g_j)$ in Equation 6.8.

For each query image, after obtaining the corresponding "code words" as described in Section 6.3, we attain its representation in the discovered concept space by substituting it in the EM iteration derived in Section 6.4.2. The only difference is that $P(r_i|z_k)$ and $P(z_k)$ are fixed to be the values we have obtained for the whole database modeling (which are obtained in the indexing phase, i.e., to determine the concept space representation of every image in the database).

In designing a region-based image mining and retrieval methodology, there are two characteristics of the region representation that must be taken into consideration:

1. The number of the segmented regions in one image is normally small.

2. Not all regions in one image are semantically relevant to a given image; some are unrelated or even non-relevant; which regions are relevant or irrelevant depends on the user's querying subjectivity.

Incorporating the "code words" corresponding to unrelated or non-relevant regions would hurt the mining or retrieval accuracy because the occurrences of these regions in one image tend to "fool" the probabilistic model such that erroneous concept representations would be generated. To address the two characteristics in image mining and retrieval explicitly, we employ the relevance feedback for the similarity measurement in the concept space. Relevance feedback has been demonstrated as great potential to capture users' querying subjectivity both in text retrieval and in image retrieval [210, 178]. Consequently, a mining and retrieval algorithm based on the relevance feedback strategy is designed to integrate the probabilistic model to deliver a more effective mining and retrieval performance.

In the algorithm, we move the query point in the "code word" token space toward the good example points (the relevant images labeled by the user) and away from the bad example points (the irrelevant images labeled by the user) such that the region representation has more supports to the probabilistic model. At the same time, the query point is expanded with the "code words" of the labeled relevant images. On the other hand, we construct a negative example "code word" vector by applying a similar vector moving strategy such that the constructed negative vector lies near the bad example points and away from the good example points. The vector moving strategy uses a form of Rocchio's formula [176]. Rocchio's formula for relevance feedback and feature expansion has proven to be one of the best iterative optimization techniques in the field of information retrieval. It is frequently used to estimate the "optimal query" in relevance feedback for sets of relevant documents D_R

and irrelevant documents D_I given by the user. The formula is

$$Q' = \alpha Q + \beta(\frac{1}{N_R} \sum_{j \in D_R} D_j) - \gamma(\frac{1}{N_I} \sum_{j \in D_I} D_j) \qquad (6.19)$$

where α, β, and γ are suitable constants; N_R and N_I are the number of documents in D_R and D_I, respectively; and Q' is the updated query of the previous query Q.

In the algorithm, based on the vector moving strategy and Rocchio's formula, in each iteration a modified query vector *pos* and a constructed negative example *neg* are computed; their representations in the discovered concept space are obtained and their similarities to each image in the database are measured through the cosine metric [12] of the corresponding vectors in the concept space, respectively. The retrieved images are ranked based on the similarity to *pos* as well as the dissimilarity to *neg*. The algorithm is described in Algorithm 9.

We use the cosine metric to compute $sim1(\bullet)$ and $sim2(\bullet)$ in Algorithm 9 because the posterior probability vectors are the basis for the similarity measure in this proposed approach. The vectors are uniform, and the value of each component in the vectors is between 0 and 1. The cosine similarity is effective and ideal for measuring the similarity for the space composed of these kinds of vectors. The experiments reported in Section 6.7 show the effectiveness of the cosine similarity measure. At the same time, we note that Algorithm 9 itself is orthogonal to the selections of similarity measure metrics. The parameters α, β, and γ in Algorithm 9 are assigned a value of 1.0 in the current implementation of the prototype system for the sake of simplicity. However, other values may be used to emphasize the different weights between good sample points and bad sample points.

6.6   Approach Analysis

It is worth comparing the proposed probabilistic model and the fitting methodology with the existing region based statistical clustering methods in the image mining and retrieval literature, such as [241, 48]. In the clustering methods, one typically associates a class variable with each image or each region in the database based on specific similarity metrics cast. One fundamental problem overlooked in such methods is that the semantic concepts of a region are typically not entirely determined by the features of the region itself; rather, they are dependent upon and affected by the contextual environment around the region in the image. In other words, a region in a different context in an image may convey a different concept. It is also noticeable that the degree of a specific region associated with several semantic concepts varies

---

Algorithm 9 A semantic concept mining based retrieval algorithm

---

Input: q, "code word" vector of the query image
Output: Images retrieved for the query image q
Method:

1: Plug q to the model to compute the vector $P(Z|q)$
2: Retrieve and rank images based on the cosine similarity measure of the vectors $P(Z|q)$ and $P(Z|g)$ of each image in the database
3: $rs = \{rel_1, rel_2, \ldots, rel_a\}$, where rel_i is a "code word" vector of each image labeled as relevant by the user on the retrieved result
4: $is = \{ire_1, ire_2, \ldots, ire_b\}$, where ire_j is a "code word" vector of each image labeled as irrelevant by the user on the retrieved result
5: $pos = \alpha q + \beta(\frac{1}{a}\sum_{i=1}^{a} rel_i) - \gamma(\frac{1}{b}\sum_{j=1}^{b} ire_j)$
6: $neg = \alpha(\frac{1}{b}\sum_{j=1}^{b} ire_j) - \gamma(\frac{1}{a}\sum_{i=1}^{a} rel_i)$
7: **for** $k = 1$ to K **do**
8:    Determine $P(z_k|pos)$ and $P(z_k|neg)$ with EM and Equation 6.18
9: **end for**
10: $n = 1$
11: **while** $n <= N$ **do**
12:    $sim1(g_n) = \frac{P(Z|pos)\bullet P(Z|g_n)}{\|P(Z|pos)\|\|P(Z|g_n)\|}$
13:    $sim2(g_n) = \frac{P(Z|neg)\bullet P(Z|g_n)}{\|P(Z|neg)\|\|P(Z|g_n)\|}$
14:    **if** $(sim1(g_n) > sim2(g_n))$ **then**
15:       $sim(g_n) = sim1(g_n) - sim2(g_n)$
16:    **else**
17:       $sim(g_n) = 0$
18:    **end if**
19:    Rank the images in the database based on $sim(g_n)$
20: **end while**

---

with different contextual region co-occurrences in an image. For example, it is likely that the *sand* "code word" conveys the concept of *beach* when it co-occurs in the context of the *water*, *sky*, and *people* "code words"; on the other hand, it becomes likely that the same *sand* "code word" conveys the concept of *African* with a high probability when it co-occurs in the context of the *plant* and *black* "code words". Wang et al [212] attempted to alleviate the effect caused by this problem by using integrated region matching to incorporate similarity between two images for all their region pairs; this matching scheme, however, is heuristic such that it is impossible for a more rigorous analysis.

The probabilistic model we have described addresses these problems quantitatively and analytically in an optimal framework. Given a region in an image the conditional probability of each concept and the conditional probability of each image in a concept are iteratively determined to fit the model representing the database as formulated in Equations 6.8 and 6.16. Since the EM technique always converges to a local optimality, from the experiments reported in Section 6.7, we have found that the local optimum is satisfactory for typical image data mining and retrieval applications. The effectiveness of this methodology in real image databases is demonstrated in the experimental analysis presented in Section 6.7. To find the global maximum is computationally intractable for a large-scale database, and the advantage of such model fitting compared to the model fitting obtained through this proposed approach is not obvious and is under further investigation.

With the proposed probabilistic model, we are able to concurrently obtain $P(z_k|r_i)$ and $P(z_k|g_j)$ such that both regions and images have an interpretation in the concept space simultaneously, while typical image clustering based approaches, such as [119], do not have this flexibility. Since in the proposed scheme, every region and/or image may be represented as a weighted sum of the components along the discovered concept axes, the proposed model acts as a factoring analysis [150], yet the same model offers important advantages, such as that each weight has a clear probabilistic meaning and the factoring is two-fold, i.e., both regions and images in the database have probabilistic representations with the discovered concepts.

Another advantage of the proposed methodology is its capability to reduce the dimensionality. The image similarity comparison is performed in a derived K-dimensional concept space Z instead of in the original M-dimensional "code word" token space R. Note that typically $K << M$, as has been demonstrated in the experiments reported in Section 6.7. The derived subspace represents the hidden semantic concepts conveyed by the regions and the images, while the noise and all the non-intrinsic information are discarded in the dimensionality reduction, which makes the semantic comparison of regions and images more effective and efficient. The coordinates in the concept space for each image as well as for each region are determined by automatic model fitting. The computation requirement in the lower-dimensional concept space is reduced as compared with that required in the original "code word" space.

Algorithm 9 integrates the posterior probability of the discovered concepts with the query expansion and the query vector moving strategy in the "code word" token space. Consequently, the accuracy of the representation of the semantic concepts of a user's query is enhanced in the "code word" token space, which also improves the accuracy of the position obtained for the query image in the concept space. Moreover, the constructed negative example *neg* improves the discriminative power of the probabilistic model. Both the similarity to the modified query representation and the dissimilarity to the constructed negative example in the concept space are employed.

6.7   Experimental Results

We have implemented the approach in a prototype system on a platform of a Pentium IV 2.0 GHz CPU and 256 MB memory. The interface of the system is shown in Figure 6.11. The following reported evaluations are performed on a general-purpose color image database containing 10,000 images from the COREL collection with 96 semantic categories. Each semantic category consists of 85–120 images. In Table 6.1, exemplar categories in the database are provided. We note that the category information in the COREL collection is only used to ground-truth the evaluation, and we do not make use of this information in the indexing, mining, and retrieval procedures. Figure 6.5 shows a few examples of the images in the database.

To evaluate the image retrieval performance, 1,500 images are randomly selected from all the categories as the query set. The relevancy of the retrieved images is subjectively examined by users. The ground truth used in the mining and retrieval experiments is the COREL category label if the query image is in the database. If the query image is a new image outside the database, users' specified relevant images in the mining and retrieval results are used to calculate the mining and retrieval accuracy statistics. Unless otherwise noted, the default results of the experiments are the averages of the top 30 returned images for each of the 1,500 queries.

In the experiments, the parameters of the image segmentation algorithm [212] are adjusted with the consideration of the balance of the depiction detail and the computation complexity such that there is an average of 8.3207 regions in each image. To determine the size of the visual token catalog, different numbers of the "code words" are selected and evaluated. The average precisions (without the query expansion and movement) within the top 20, 30, and 50 images, denoted as P(20), P(30), and P(50), respectively, are shown in Figure 6.6. It indicates that the general trend is that the larger the visual token catalog size, the higher the mining and retrieval accuracy. However, a larger visual token catalog size means a larger number of image feature vectors, which implies a higher computation complexity in the process of the

Table 6.1: Examples of the 96 categories and their descriptions. Reprint from [243] ©2007 IEEE Signal Processing Society Press.

ID	Category description
1	reptile, animal, rock
2	Britain, royal events, queen, prince, princess
3	Africa, people, landscape, animal
4	European, historical building, church
5	woman, fashion, model, face, cloth
6	hawk, sky
7	New York City, skyscrapers, skyline
8	mountain, landscape
9	antique, craft
10	Easter egg, decoration, indoor, man-made
11	waterfall, river, outdoor
12	poker cards
13	beach, vacation, sea shore, people
14	castle, grass, sky
15	cuisine, food, indoor
16	architecture, building, historical building
..	......

FIGURE 6.5: Sample images in the database. The images in each column are assigned to one category. From left to right, the categories are Africa rural area, historical building, waterfalls, British royal event, and model portrait, respectively.

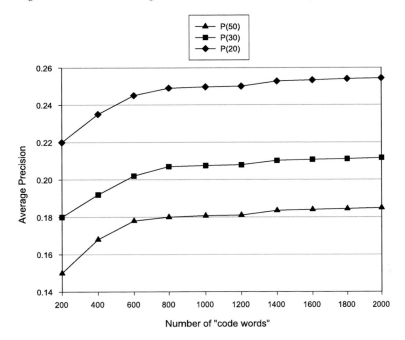

FIGURE 6.6: Average precision (without the query expansion and movement) for different sizes of the visual token catalog. Reprint from [243] ©2007 IEEE Signal Processing Society Press and from [240] ©2004 IEEE Computer Society Press.

hidden semantic concept discovery. Also, a larger visual token catalog leads to a larger storage space. Therefore, we use 800 as the number of the "code words", which corresponds to the first turning point in Figure 6.6. Since there are a total of 83,307 regions in the database, on average each "code word" represents 104.13 regions.

Applying the method of estimating the number of the hidden concepts described in Section 6.4.3, the number of the concepts is determined to be 132. Performing the EM model fitting, we have obtained the conditional probability of each "code word" to every concept, i.e., $P(r_i|z_k)$. Manual examination of the visual content of the region sets corresponding to the top 10 highest "code words" in every semantic concept reveals that these discovered concepts indicate semantic interpretations, such as "people", "building", "outdoor scenery", "plant", and "automotive race". Figure 6.7 shows several exemplar concepts discovered and the top regions corresponding to $P(r_i|z_k)$ obtained.

In terms of the computational complexity, despite the iterative nature of EM, the computing time for the model fitting at $K = 132$ is acceptable (less than 1 second). The average number of iterations upon convergence for one

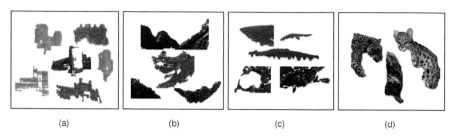

FIGURE 6.7: The regions with the top $P(r_i|z_k)$ to the different concepts discovered. (a) "castle"; (b) "mountain"; (c) "meadow and plant"; (d) "cat". Reprint from [243] ©2007 IEEE Signal Processing Society Press.

FIGURE 6.8: Illustration of one query image in the "code word" space. (a) Image Im; (b) "code word" representation. Reprint from [243] ©2007 IEEE Signal Processing Society Press.

image is less than 5.

We give an example for discussion. Figure 6.8 shows one image, Im, belonging to the "medieval building" category in the database. Im (i.e., Figure 6.8(a)) has 6 "code words" associated. Each "code word" is presented using a unique color graphically in Figure 6.8(b). For the sake of discussion, the indices for these "code words" are assigned to be 1–6, respectively.

Figure 6.9 shows the $P(z_k|r_i, Im)$ for each "code word" r_i (represented as a different color) and the posterior probability $P(z_k|Im)$ after the first iteration and the last iteration in the course of the EM model fitting. Here the 4 concepts with highest $P(z_k|Im)$ are shown. From left to right in Figure 6.9, they represent "plant", "castle", "cat", and "mountain", respectively, interpreted through manual examination. As is seen in the figure, the "castle" concept has indeed the highest weight after the first iteration; nevertheless, the other three concepts still account for more than half of the probability. The probability distribution changes after several EM iterations, since the proposed probabilistic model incorporates co-occurrence patterns between the "code words"; i.e., $P(z_k|r_i)$ is not only related to one "code word" (r_i) but is also related to all the co-occurring "code words" in the image. For example, although "code word" 2, which accounts for "meadow", has higher fitness in the concept "plant" after the first iteration, the context of the other regions

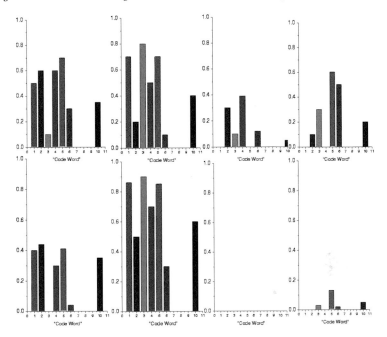

FIGURE 6.9: $P(z_k|r_i, Im)$ (each color column for a "code word") and $P(z_k|Im)$ (rightmost column in each bar plot) for image Im for the four concept classes (semantically related to "plant", "castle", "cat", and "mountain", from left to right, respectively) after the first iteration (first row) and the last iteration (second row). Reprint from [243] ©2007 IEEE Signal Processing Society Press.

in image Im increases the probability that this "code word" is related to the concept "castle" and decreases its probability related to "plant" as well.

Figure 6.10 shows the similar plot to Figure 6.9 except that we apply the relevance feedback based query expansion and moving strategy to image Im as described in the Algorithm 9. The "code word" vector of image Im is expanded to contain 10 "code words". Compared with Figure 6.9, it is clear that with the expansion of the relevant "code words" to Im and the query moving strategy toward the relevant image set, the posterior probabilities favoring the concept "castle" increase while the posterior probabilities favoring other concepts decrease substantially, resulting in an improved mining and retrieval precision, accordingly.

To show the effectiveness of the probabilistic model in image mining and retrieval, we have compared the accuracy of this methodology with that of UFM [47] proposed by Chen and Wang. UFM is a method based on the

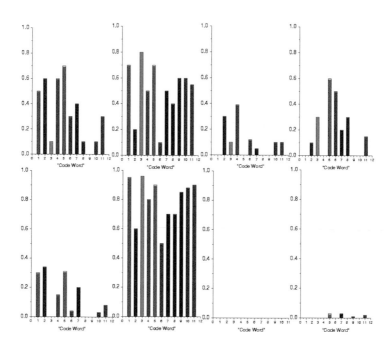

FIGURE 6.10: The similar plot to Figure 6.9 with the application of the query expansion and moving strategy. Reprint from [243] ©2007 IEEE Signal Processing Society Press.

fuzzified region representation to build region-to-region similarity measures for image retrieval; it is an improvement of their early work SIMPLIcity [212]. The reasons why we compare this proposed approach with UFM are: (1) the UFM system is available to us; and (2) UFM reflects the performance of the state-of-the-art image mining and retrieval performance. In addition, the same image segmentation and feature extraction methods are used in UFM such that a fair comparison on the performance between the two systems is ensured. Figure 6.11 shows the top 16 retrieved images by the prototype system and as well as by UFM, respectively, using image Im as a query.

More systematic comparison results on the 1,500 query image set are reported in Figure 6.12. Two versions of the prototype (one with the query expansion and moving strategy and the other without) and UFM are evaluated. It is demonstrated that the performances of the probabilistic model in both versions of the prototype have higher overall precisions than that of UFM, and the query expansion and moving strategy with the interaction of the constructed negative examples boost the mining and retrieval accuracy significantly.

6.8 Summary

In this chapter we have presented an approach to image data mining and retrieval based on automatically discovering the hidden semantic concepts in the database. The main contributions of the work described in this chapter are the identification of the problems existing in most region-based image mining and retrieval methods —- unreliable region evidence in semantic contents and the development of a promising hidden semantics concept discovery technique to solve the problems. Performing image segmentation with multiple features and developing an SOM based quantization method to generate a visual token catalog, a uniform and sparse region-based representation scheme is obtained. On the basis of this representation, a probabilistic model of the image database is defined. The model assumes that the regions, hidden semantic concepts, and images are random variables and the objective is to discover concept distributions with samples from the (region, image) pair distributions. Based on this model, the EM method is applied to derive an iterative procedure to discover the hidden semantic concepts in the database. An elaborated relevance feedback based mining and retrieval algorithm is designed to support the model and to improve the mining and retrieval accuracy. The image querying is performed by integrating the posterior probabilities of the transformed images for the discovered semantic concepts. Supported by a solid statistical foundation, this approach enables mining and retrieval by higher-order semantic indicants that are more reliable, hence improving the

(a)

(b)

FIGURE 6.11: Retrieval performance comparisons between UFM and the prototype system using image Im in Figure 6.8 as the query. (a) Images returned by UFM (9 of the 16 images are relevant). (b) Images returned by the prototype system (14 of the 16 images are relevant).

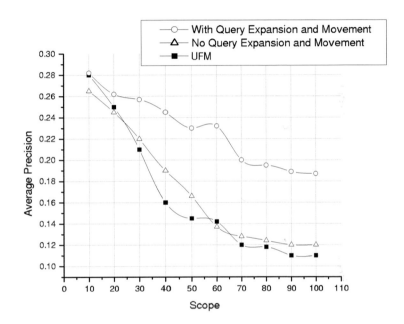

FIGURE 6.12: Average precision comparisons between the two versions of the prototype and UFM. Reprint from [243] ©2007 IEEE Signal Processing Society Press and from [240] ©2004 IEEE Computer Society Press.

mining and retrieval accuracy. The experimental evaluations on a database of 10,000 general-purpose images demonstrate the effectiveness and the promise of the approach in general image data mining and retrieval.

Chapter 7

A Multimodal Approach to Image Data Mining and Concept Discovery

7.1 Introduction

This chapter gives an example on multimedia data mining by addressing the automatic image annotation problem and its application to multimodal image data mining and retrieval. Specifically, in this chapter, we propose a probabilistic semantic model in which the visual features and the textual words are connected via a hidden layer which constitutes the semantic concepts to be discovered to explicitly exploit the synergy between the two modalities; the association of visual features and the textual words is determined in a Bayesian framework such that the confidence of the association can be provided; and extensive evaluations on a large-scale, visually and semantically diverse image collection crawled from the Web are reported to evaluate the prototype system based on the model. In the proposed probabilistic model, a hidden concept layer which connects the visual features and the word layer is discovered by fitting a generative model to the training images and annotation words. An Expectation-Maximization (EM) based iterative learning procedure is developed to determine the conditional probabilities of the visual features and the textual words given a hidden concept class. Based on the discovered hidden concept layer and the corresponding conditional probabilities, the image annotation and the text-to-image retrieval are performed using the Bayesian framework. The evaluations of the prototype system on 17,000 images and 7,736 automatically extracted annotation words from the crawled Web pages for multimodal image data mining and retrieval have indicated that the model and the framework are superior to a state-of-the-art peer system in the literature.

The rest of the chapter is organized as follows: Section 7.2 introduces the motivations to this work and outlines the main contributions of this work. Section 7.3 discusses the related work on image annotation and multimodal image mining and retrieval. In Section 7.4 the proposed probabilistic semantic model and the EM based learning procedure are described. Section 7.5 presents the Bayesian framework developed to support the multimodal image data mining and retrieval. The acquisition of the training and testing data collected from the Web, and the experiments to evaluate the proposed approach against a state-of-the-art peer system in several aspects, are reported in Section 7.6. Finally, this chapter is concluded in Section 7.7.

7.2 Background

Efficient access to multimedia database requires the ability to search and organize multimedia information. In traditional image retrieval, users have to provide examples of images that they are looking for. Similar images are found based on the match of image features. Even though there have been many studies on this traditional image retrieval paradigm, empirical studies have shown that using image features solely to find similar images is usually insufficient due to the notorious *semantic gap* between low-level features and high-level semantic concepts [192]. As a step further to reduce this gap, region based features (describing object level features), instead of raw features of whole image, to represent the visual content of an image are proposed [37, 212, 47].

On the other hand, it is well-observed that often imagery does not exist in isolation; instead, typically there is rich collateral information co-existing with image data in many applications. Examples include the Web, many domain-archived image databases (in which there are annotations to images), and even consumer photo collections. In order to further reduce the semantic gap, recently multimodal approaches to image data mining and retrieval have been proposed in the literature [251] to explicitly exploit the redundancy co-existing in the collateral information to the images. In addition to the improved mining and retrieval accuracy, a benefit for the multimodal approaches is the added querying modalities. Users can query an image database either by imagery, by a collateral information modality (e.g., text), or by any combination.

In this chapter, we propose a probabilistic semantic model and the corresponding learning procedure to address the problem of automatic image annotation and show its application to multimodal image data mining and retrieval. Specifically, we use the proposed probabilistic semantic model to explicitly exploit the synergy between the different modalities of the imagery and the collateral information. In this work, we only focus on a specific collateral modality — text. The model may be generalized to incorporate other collateral modalities. Consequently, the synergy here is explicitly represented as a hidden layer between the imagery and the text modalities. This hidden layer constitutes the concepts to be discovered through a probabilistic framework such that the confidence of the association can be provided. An Expectation-Maximization (EM) based iterative learning procedure is developed to determine the conditional probabilities of the visual features and the words given a hidden concept class. Based on the discovered hidden concept layer and the corresponding conditional probabilities, the image-to-text and text-to-image retrievals are performed in a Bayesian framework.

In recent image data mining and retrieval literature, COREL data have been extensively used to evaluate the performance [14, 70, 75, 136]. It has been argued [217] that the COREL data are much easier to annotate and retrieve due to their small number of concepts and small variations of the visual content. In addition, the relative small number (1,000 to 5,000) of the training images and test images typically used in the literature further makes the problem easier and the evaluation less convictive. In order to truly capture the difficulties in real scenarios such as Web image data mining and retrieval and to demonstrate the robustness and the promise of the proposed model and the framework in these challenging applications, we have evaluated the prototype system on a collection of 17,000 images with the automatically extracted textual annotations from various crawled Web pages. We have shown that the proposed model and framework work well on this scale of a very noisy image dataset and substantially outperform the state-of-the-art peer system MBRM [75].

The specific contributions of this work include:

1. We propose a probabilistic semantic model in which the visual features and textual words are connected via a hidden layer to constitute the concepts to be discovered to explicitly exploit the synergy between the two modalities. An EM based learning procedure is developed to fit the model to the two modalities.

2. The association of visual features and textual words is determined in a Bayesian framework such that the confidence of the association can be provided.

3. Extensive evaluations on a large-scale collection of visually and semantically diverse images crawled from the Web are performed to evaluate the prototype system based on the model and the framework. The experimental results demonstrate the superiority and the promise of the approach.

7.3   Related Work

A number of approaches have been proposed in the literature on automatic image annotation [14, 70, 75, 136]. Different models and machine learning techniques are developed to learn the correlation between image features and textual words from the examples of annotated images and then apply the learned correlation to predict words for unseen images. The co-occurrence model [156] collects the co-occurrence counts between words and image features and uses them to predict annotated words for images. Barnard and Duygulu et al [14, 70] improved the co-occurrence model by utilizing machine translation models. The models are correspondence extensions to Hofmann et al's hierarchical clustering aspect model [102, 103, 101], and incorporate multi-modality information. The models consider image annotation as a process of translation from "visual language" to text and collect the co-occurrence information by the estimation of the translation probabilities. The correspondence between *blobs* and words are learned by using statistical translation models. As noted by the authors [14], the performance of the models is strongly affected by the quality of image segmentation. More sophisticated graphical models, such as Latent Dirichlet Allocation (LDA) [22] and correspondence LDA, have also been applied to the image annotation problem recently [21]. Specific reviews on using the graphical models for multimedia data mining including image annotation are given in Section 3.6.

Another way to address automatic image annotation is to apply classification approaches. The classification approaches treat each annotated word (or each semantic category) as an independent class and create a different image classification model for every word (or category). One representative work of these approaches is the automatic linguistic indexing of pictures (ALIPS) [136]. In ALIPS, the training image set is assumed well classified and each category is modeled by using 2D multi-resolution hidden Markov models. The image annotation is based on the nearest-neighbor classification and word occurrence counting, while the correspondence between the visual content and the annotation words is not exploited. In addition, the assumption made in ALIPS that the annotation words are semantically exclusive is not valid in nature.

Recently, relevance language models [75] have been successfully applied to automatic image annotation. The essential idea is to first find annotated images that are similar to a test image and then use the words shared by the annotations of the similar images to annotate the test image. One model in this category is the Multiple-Bernoulli Relevance Model (MBRM) [75], which is based on the Continuous-space Relevance Model (CRM) [134]. In MBRM, the word probabilities are estimated using a multiple Bernoulli model and the image block feature probabilities are estimated using a non-parametric kernel density estimate. The reported experiments show that the MBRM model outperforms the previous CRM model, which assumes that annotation words for any given image follow a multinomial distribution and applies image segmentation to obtain blobs for annotation.

It has been noted that in many cases both images and word-based documents are of interest to users' querying needs, such as in the Web search environment. In these scenarios, multimodal image data mining and retrieval, i.e., leveraging the collected textual information to improve image mining and retrieval and to enhance users' querying modalities, are proven to be very promising. Studies have been reported on this problem. Chang et al [40] have applied the Bayes Point Machine to associate words and images to support multimodal image mining and retrieval. In [252], latent semantic indexing is used together with both textual and visual features to extract the underlying semantic structures of Web documents. Improvement of the mining and retrieval performance is reported, attributing to the synergy of both modalities.

7.4 Probabilistic Semantic Model

To achieve automatic image annotation as well as multimodal image data mining and retrieval, a probabilistic semantic model is proposed for the training imagery and the associated textual word annotation dataset. The probabilistic semantic model is developed by the EM technique to determine the hidden layer connecting image features and textual words, which constitutes the semantic concepts to be discovered to explicitly exploit the synergy between the imagery and text.

7.4.1 Probabilistically Annotated Image Model

First, a word about notation: $f_i, i \in [1, N]$ denotes the visual feature vector of images in the training database, where N is the size of the image database. $w^j, j \in [1, M]$ denotes the distinct textual words in the training annotation word set, where M is the size of annotation vocabulary in the training database.

In the probabilistic model, we assume the visual features of images in the database, $f_i = [f_i^1, f_i^2, \ldots, f_i^L], i \in [1, N]$, are known i.i.d. samples from an unknown distribution. The dimension of the visual feature is L. We also assume that the specific visual feature annotation word pairs $(f_i, w^j), i \in [1, N], j \in [1, M]$ are known i.i.d. samples from an unknown distribution. Furthermore, we assume that these samples are associated with an unobserved *semantic concept* variable $z \in Z = \{z_1, \ldots, z_K\}$. Each observation of one visual feature $f \in F = \{f_1, f_2, \ldots, f_N\}$ belongs to one or more concept classes z_k, and each observation of one word $w \in V = \{w^1, w^2, \ldots, w^M\}$ in one image f_i belongs to one concept class. To simplify the model, we have two more assumptions. First, the observation pairs (f_i, w^j) are generated independently. Second, the pairs of random variables (f_i, w^j) are conditionally independent given the respective hidden concept z_k,

$$P(f_i, w^j | z_k) = p_{\mathcal{F}}(f_i | z_k) P_{\mathcal{V}}(w^j | z_k) \tag{7.1}$$

The visual feature and word distribution are treated as a randomized data generation process, described as follows:

- Choose a concept with probability $P_{\mathcal{Z}}(z_k)$;

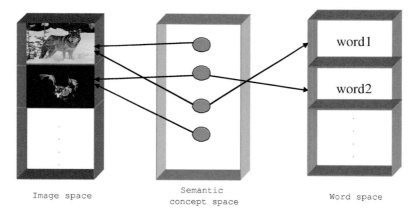

Image space   Semantic   Word space
                  concept space

FIGURE 7.1: Graphic representation of the model proposed for the randomized data generation for exploiting the synergy between imagery and text.

- Select a visual feature $f_i \in F$ with probability $P_{\mathcal{F}}(f_i|z_k)$; and

- Select a textual word $w^j \in V$ with probability $P_{\mathcal{V}}(w^j|z_k)$.

As a result, one obtains an observed pair (f_i, w^j), while the concept variable z_k is discarded. The graphic representation of this model is depicted in Figure 7.1.

Translating this process into a joint probability model results in the expression

$$P(f_i, w^j) = P(w^j)P(f_i|w^j)$$

$$= P(w^j) \sum_{k=1}^{K} P_{\mathcal{F}}(f_i|z_k)P(z_k|w^j) \qquad (7.2)$$

Inverting the conditional probability $P(z_k|w^j)$ in Equation 7.2 with the application of Bayes' rule results in

$$P(f_i, w^j) = \sum_{k=1}^{K} P_{\mathcal{Z}}(z_k)P_{\mathcal{F}}(f_i|z_k)P_{\mathcal{V}}(w^j|z_k) \qquad (7.3)$$

The mixture of Gaussian [60] is assumed for the feature-concept conditional probability $P_{\mathcal{F}}(\bullet|Z)$. In other words, the visual features are generated from K Gaussian distributions, each one corresponding to a z_k. For a specific semantic concept variable z_k, the conditional pdf of visual feature f_i is

$$p_{\mathcal{F}}(f_i|z_k) = \frac{1}{(2\pi)^{L/2}|\sum_k|^{1/2}} e^{-\frac{1}{2}(f_i-\mu_k)^T \sum_k^{-1}(f_i-\mu_k)} \qquad (7.4)$$

where $\sum_k$ and μ_k are the covariance matrix and mean of the visual features belonging to z_k, respectively. The word-concept conditional probabilities $P_V(\bullet|Z)$, i.e., $P_V(w^j|z_k)$ for $k \in [1, K]$, are estimated through fitting the probabilistic model to the training set.

Following the likelihood principle, one determines $P_{\mathcal{F}}(f_i|z_k)$ by the maximization of the log-likelihood function

$$\log \prod_{i=1}^{N} p_{\mathcal{F}}(f_i|Z)^{u_i} = \sum_{i=1}^{N} u_i \log(\sum_{k=1}^{K} P_{\mathcal{Z}}(z_k)p_{\mathcal{F}}(f_i|z_k)) \qquad (7.5)$$

where u_i is the number of the annotation words for image f_i. Similarly, $P_{\mathcal{Z}}(z_k)$ and $P_V(w^j|z_k)$ can be determined by the maximization of the log-likelihood function

$$\mathcal{L} = \log P(F, V) = \sum_{i=1}^{N} \sum_{j=1}^{M} n(w_i^j) \log P(f_i, w^j) \qquad (7.6)$$

where $n(w_i^j)$ denotes the weight of annotation word w^j, i.e., the occurrence frequency, for image f_i.

7.4.2 EM Based Procedure for Model Fitting

From Equations 7.5, 7.6, and 7.2, we derive that the model is a statistical mixture model [150], which can be resolved by applying the EM technique [58]. The EM alternates in two steps: (i) an expectation (E) step where the posterior probabilities are computed for the hidden variable z_k, based on the current estimates of the parameters; and (ii) a maximization (M) step, where parameters are updated to maximize the expectation of the complete-data likelihood $\log P(F, V, Z)$ given the posterior probabilities computed in the previous E-step. Thus, the probabilities can be iteratively determined by fitting the model to the training image database and the associated annotations.

Applying Bayes' rule to Equation 7.3, we determine the posterior probability for z_k under f_i and (f_i, w^j):

$$p(z_k|f_i) = \frac{P_{\mathcal{Z}}(z_k)p_{\mathcal{F}}(f_i|z_k)}{\sum_{t=1}^{K} P_{\mathcal{Z}}(z_t)p_{\mathcal{F}}(f_i|z_t)} \qquad (7.7)$$

$$P(z_k|f_i, w^j) = \frac{P_{\mathcal{Z}}(z_k)P_{\mathcal{Z}}(f_i|z_k)P_V(w^j|z_k)}{\sum_{t=1}^{K} P_{\mathcal{Z}}(z_t)P_{\mathcal{F}}(f_i|z_t)P_V(w^j|z_t)} \qquad (7.8)$$

The expectation of the complete-data likelihood $\log P(F, V, Z)$ for the estimated $P(Z|F, V)$ derived from Equation 7.8 is

$$\sum_{(i,j)=1}^{K} \sum_{i=1}^{N} \sum_{j=1}^{M} n(w_i^j) \log [P_{\mathcal{Z}}(z_{i,j})p_{\mathcal{F}}(f_i|z_{i,j})P_V(w^j|z_{i,j})]P(Z|F, V) \qquad (7.9)$$

where

$$P(Z|F,V) = \prod_{s=1}^{N} \prod_{t=1}^{M} P(z_{s,t}|f_s, w^t)$$

In Equation 7.9 the notation $z_{i,j}$ is the concept variable that associates with the feature-word pair (f_i, w^j). In other words, (f_i, w^j) belongs to concept z_t where $t = (i,j)$.

Similarly, the expectation of the likelihood $\log P(F, Z)$ for the estimated $P(Z|F)$ derived from Equation 7.7 is

$$\sum_{k=1}^{K} \sum_{i=1}^{N} \log(P_Z(z_k)p_F(f_i|z_k))p(z_k|f_i) \qquad (7.10)$$

Maximizing Equations 7.9 and 7.10 with Lagrange multipliers to $P_Z(z_l)$, $p_F(f_u|z_l)$, and $P_V(w^v|z_l)$, respectively, under the following normalization constraints

$$\sum_{k=1}^{K} P_Z(z_k) = 1, \sum_{k=1}^{K} P(z_k|f_i, w^j) = 1 \qquad (7.11)$$

for any f_i, w^j, and z_l, the parameters are determined as

$$\mu_k = \frac{\sum_{i=1}^{N} u_i f_i p(z_k|f_i)}{\sum_{s=1}^{N} u_s p(z_k|f_s)} \qquad (7.12)$$

$$\sum_k = \frac{\sum_{i=1}^{N} u_i p(z_k|f_i)(f_i - \mu_k)(f_i - \mu_k)^T}{\sum_{s=1}^{N} u_s p(z_k|f_s)} \qquad (7.13)$$

$$P_Z(z_k) = \frac{\sum_{j=1}^{M} \sum_{i=1}^{N} u(w_i^j)P(z_k|f_i, w^j)}{\sum_{j=1}^{M} \sum_{i=1}^{N} n(w_i^j)} \qquad (7.14)$$

$$P_V(w^j|z_k) = \frac{\sum_{i=1}^{N} n(w_i^j)P(z_k|f_i, w^j)}{\sum_{u=1}^{M} \sum_{v=1}^{N} n(w_v^u)P(z_k|f_v, w^u)} \qquad (7.15)$$

Alternating Equations 7.7 and 7.8 with Equations 7.12–7.15 defines a convergent procedure to a local maximum of the expectation in Equations 7.9 and 7.10.

7.4.3 Estimating the Number of Concepts

The number of concepts, K, must be determined in advance for the EM model fitting. Ideally, we intend to select the value of K that best agrees with the number of the semantic classes in the training set. One readily available notion of the fitting goodness is the log-likelihood. Given this indicator, we can apply the Minimum Description Length (MDL) principle [175] to select among values of K. This can be done as follows [175]: choose K to maximize

$$\log(P(F, V)) - \frac{m_K}{2} \log(MN) \qquad (7.16)$$

where the first term is expressed in Equation 7.6 and m_K is the number of free parameters needed for a model with K mixture components. In our probabilistic model, we have

$$m_K = (K - 1) + K(M - 1) + K(N - 1) + L^2 = K(M + N - 1) + L^2 - 1$$

As a consequence of this principle, when models with different values of K fit the data equally well, the simpler model is selected. In the experimental database reported in Section 7.6, K is determined through maximizing Equation 7.16.

7.5 Model Based Image Annotation and Multimodal Image Mining and Retrieval

After the EM based iterative procedure converges, the model fitting to the training set is obtained. The image annotation and multimodal image mining and retrieval are conducted in a Bayesian framework with the determined $P_{\mathcal{Z}}(z_k)$, $p_{\mathcal{F}}(f_i|z_k)$, and $P_{\mathcal{V}}(w^j|z_k)$.

7.5.1 Image Annotation and Image-to-Text Querying

The objective of image annotation is to return words which best reflect the semantics of the visual content of images. In this proposed approach, we use a joint distribution to model the probability of an event that a word w^j belonging to semantic concept z_k is an annotation word of image f_i. Observing Equation 7.1, the joint probability is

$$P(w^j, z_k, f_i) = P_{\mathcal{Z}}(Z_k) p_{\mathcal{F}}(f_i|z_k) P_{\mathcal{V}}(w^j|z_k) \qquad (7.17)$$

Through applying Bayes' law and the integration over $P_{\mathcal{Z}}(z_k)$, we obtain the following expression:

$$
\begin{aligned}
P(w^j|f_i) &= \int P_{\mathcal{V}}(w^j|z) p(z|f_i) dz \\
&= \int P_{\mathcal{V}}(w^j|z) \frac{p_{\mathcal{F}}(f_i|z) P(z)}{p(f_i)} dz \\
&= E_z\{ \frac{P_{\mathcal{V}}(w^j|z) p_{\mathcal{F}}(f_i|z)}{p(f_i)} \}
\end{aligned} \qquad (7.18)
$$

where

$$p(f_i) = \int p_{\mathcal{F}}(f_i|z)P_{\mathcal{Z}}(z)dz = E_z\{p_{\mathcal{F}}(f_i|z)\} \qquad (7.19)$$

In the above equations $E_z\{\bullet\}$ denotes the expectation over $P(z_k)$, the probability of semantic concept variables. Equation 7.18 provides a principled way to determine the probability of word w^j for annotating image f_i. With the combination of Equations 7.18 and 7.19, the automatic image annotation can be solved fully in the Bayesian framework.

In practice, we derive an approximation of the expectation in Equation 7.18 by utilizing the Monte Carlo sampling [79] technique. Applying Monte Carlo integration to Equation 7.18 derives

$$P(w^j|f_i) \approx \frac{\sum_{k=1}^{K} P_{\mathcal{V}}(w^j|z_k)p_{\mathcal{F}}(f_i|z_k)}{\sum_{h=1}^{K} p_{\mathcal{F}}(f_i|z_h)}$$

$$= \sum_{k=1}^{K} P_{\mathcal{V}}(w^j|z_k)x_k \qquad (7.20)$$

where $x_k = \frac{p_{\mathcal{F}}(f_i|z_k)}{\sum_{h=1}^{K} p_{\mathcal{F}}(f_i|z_h)}$. The words with the top highest $P(w^j|f_i)$ are returned to annotate the image. Given this image annotation scheme, the image-to-text querying may be performed by retrieving documents for the returned words based on the traditional text retrieval techniques.

7.5.2   Text-to-Image Querying

The traditional text-based image retrieval systems, e.g., Google image search, solely use textual information to index images. It is well-known that this approach fails to achieve satisfactory image retrieval, which actually has motivated the content based image indexing research. Based on the model obtained in Section 7.4 to explicitly exploit the synergy between imagery and text, we here develop an alternative and much more effective approach using the Bayesian framework to image data mining and retrieval given a text query.

Similar to the derivation in Section 7.5.1, we retrieve images for word queries by determining the conditional probability $P(f_i|w^j)$:

$$P(f_i|w^j) = \int P_{\mathcal{F}}(f_i|z)P(z|w^j)dz$$

$$= \int P_{\mathcal{V}}(w^j|z)\frac{p_{\mathcal{F}}(f_i|z)P(z)}{P(w^j)}dz$$

$$= E_z\{\frac{P_{\mathcal{V}}(w^j|z)p_{\mathcal{F}}(f_i|z)}{P(w^j)}\} \qquad (7.21)$$

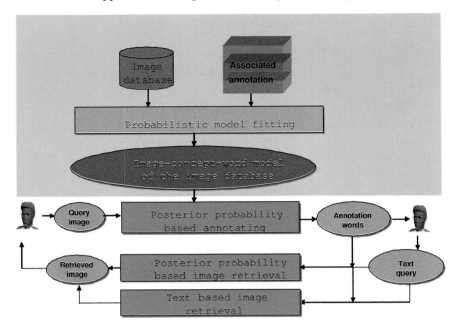

FIGURE 7.2: The architecture of the prototype system.

The expectation can be estimated as follows:

$$P(f_i|w^j) \approx \frac{\sum_{k=1}^{K} P_\mathcal{V}(w^j|z_k)p_\mathcal{F}(f_i|z_k)}{\sum_{h=1}^{K} P_\mathcal{V}(w^j|z_h)}$$

$$= \sum_{k=1}^{K} p_\mathcal{F}(f_i|z_k)y_k \tag{7.22}$$

where $y_k = \frac{P_\mathcal{V}(w^j|z_k)}{\sum_h P_\mathcal{V}(w^j|z_h)}$. The images in the database with the top highest $P(f_i|w^j)$ are returned as the querying result for each query word.

7.6 Experiments

We have implemented the approach in a prototype system. The architecture of the prototype system is illustrated in Figure 7.2. The system supports both image-to-text (i.e., image annotation) and text-to-image queryings.

{people (6), mountain(6), sky(5)}

FIGURE 7.3: An example of image and annotation word pairs in the generated database. The number following each word is the corresponding weight of the word.

7.6.1 Dataset and Feature Sets

It has been noted that the datasets used in the recent automatic image annotation systems [14, 70, 75, 136] fail to capture the difficulties inherent in many real image databases. Two issues are taken into the consideration in the design of the experiments reported in this section. First, the commonly used COREL database is much easier for image annotation and retrieval due to its limited semantics conveyed and small variations of the visual content. Second, the typical small scales of the datasets reported in the recent literature are far away from being realistic in all the real-world applications. To address these issues, we decide not to use the COREL database in the evaluation of the prototype system; instead, we evaluate the system on a collection of a large-scale real-world dataset automatically crawled from the Web. The web pages crawled are from the Yahoo! photos website; then the images and the surrounding text describing the images' content are extracted from the blocks containing the images by using the VIPS algorithm [32]. The surrounding text is processed using the standard text processing techniques to obtain the annotation words. Apart from the images and the annotation words, the weight of each annotation word for the images is computed by using a scheme incorporating TF, IDF, and the tag information in VIPS, and is normalized to range (0,10]. The image-annotation word pairs are stemmed and manually cleaned before using as the training database for the model fitting and testing. The data collection consists of 17,000 images and 7,736 stemmed annotation words. Among them, 12,000 images are used as the training set and the remaining 5,000 images are used for the testing purpose. Compared with images in COREL, the images in this set are more diverse both on semantics and on visual appearance, which reflects the true nature of image search in many real applications. Figure 7.3 shows an image example with the associated annotation words in the generated database.

The focus of this chapter is not on image feature selection and the proposed approach is independent of any visual features. For implementation simplicity and easy comparison purposes, similar features used in [75] are used in the prototype system. Specifically, a visual feature is a 36-dimensional vector, consisting of 24 color features (auto correlogram computed over 8 quantized colors and 3 Manhattan Distances) and 12 texture features (Gabor energy computed over 3 scales and 4 orientations).

7.6.2 Evaluation Metrics

To evaluate the effectiveness and the promise of the prototype system for multimodal image data mining and retrieval, the following performance measures are defined:

- Hit-Rate3 (HR3): the average rate of at least one word in the ground truth of a test image is returned in the top 3 returned words for the test set.

- Complete-Length (CL): the average minimum length of the returned words which contain all the ground truth words for a test image for the test set.

- Single-Word-Query-Precision (SWQP(n)): the average rate of the relevant images (here "relevant" means that the ground truth annotation of this image contains the query word) in the top n returned images for a single word query for the test set.

HR3 and CL measure the accuracy of image annotation (or the image-to-text querying); the higher the HR3, and/or the lower the CL, the better the annotation accuracy. SWQP(n) measures the precision of the text-to-image querying; the higher the SWQP(n), the better the text-to-image querying precision.

Furthermore, we also measure the image annotation performance by using the annotation recall and precision defined in [75]. $recall = \frac{B}{C}$ and $precision = \frac{B}{A}$, where A is the number of the images automatically annotated with a given word in the top 10 returned word list; B is the number of the images correctly annotated with that word in the top-10-returned-word list; and C is the number of the images having that word in the ground truth annotation. An ideal image annotation system would have a high average annotation recall and annotation precision simultaneously.

7.6.3 Results of Automatic Image Annotation

The interface of the prototype system for automatic image annotation is shown in Figure 7.4. In this system, words and their confidence scores (conditional probabilities) are returned to annotate images upon users' querying.

Applying the method of estimating the number of the hidden concepts described in Section 7.4.3 to the training set, the number of the concepts is determined to be 262. Compared with the number of the images in the

Table 7.1: Comparisons between the examples of the automatic annotations generated by the proposed prototype system and MBRM. Reprint from [246] ©2006 Springer-Verlag Press and from [245] ©2005 IEEE Computer Society Press.

Systems	MBRM	Our Prototype
	animal, water, wolf, house, tiger	wolf, winter, wild, animal, stone
	male-face, hair, people, bear, sky	male-face, people, hair, man, monologue
	bird, grass, leopard, sail, cuckoo	bird, cuckoo, yellow, sand, sky
	flower, red, tree, meadow, outdoor	flower, red, azalea, leaf, landscape
	desert, beach, mummy, building, church	pyramid, Egypt, desert, mummy, beach

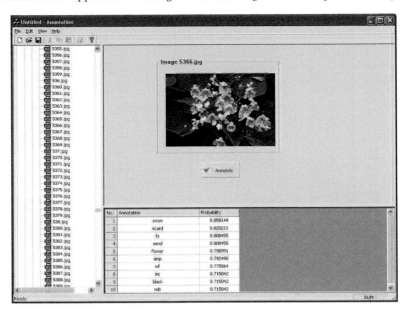

FIGURE 7.4: The interface of the automatic image annotation prototype.

training set, 12,000, and the number of the stemmed and cleaned annotation words, 7,736, the number of the semantic concept variables is far less. In terms of the computational complexity, the model fitting is computation-intensive; it takes 45 hours to fit the model to the training set on a Pentium IV 2.3 GHz computer with 1 GB memory. Fortunately, this process is performed offline and only once. For online image annotation and single-word image querying, the response time is acceptable (less than 1 second).

To show the effectiveness and the promise of the probabilistic model in image annotation, we have compared the accuracy of the proposed method with that of MBRM [75]. In MBRM, the word probabilities are estimated using a multiple Bernoulli model, and no association layer between visual features and words is used. We compare the proposed approach with MBRM because MBRM reflects the performance of the state-of-the-art automatic image annotation research. In addition, since the same image visual features are used in MBRM, a fair comparison of the performance is expected. Table 7.1 shows examples of the automatic annotation obtained by the proposed prototype system and MBRM on the test image set. Here the top 5 words (according to probability) are taken as the automatic annotation of an image. The performance comparison demonstrated in the table clearly indicates that the proposed system performs better than MBRM.

The systematic evaluation results are shown for the test set in Table 7.2. The results are reported for all (7,736) words in the database. The proposed approach clearly outperforms MBRM. As is shown, the average recall im-

Table 7.2: Performance comparison on the task of automatic image annotation on the test set. Reprint from [246] ©2006 Springer-Verlag Press and from [245] ©2005 IEEE Computer Society Press.

Models	MBRM	Proposed Model
HR3	0.56	0.83
CL	1265	574
#words with $recall > 0$	3295	6078
Results on all 7736 words		
Average Per-word Recall	0.19	0.28
Average Per-word Precision	0.16	0.27

proves by 48% and the average precision improves by 69%. The multiple Bernoulli generation of the words in MBRM is artificial and the association of the words and features is noisy. On the contrary, in the proposed model no explicit word distribution is assumed, and the synergy between the visual features and the words exploited by the hidden concept variables reduces the noise substantially. We believe that these reasons account for the better performance of the proposed approach. We note that certain returned words with top rank from the proposed system for a given image query are found semantically relevant by subjective examinations, although they are not contained in the ground truth annotation of the image. We did not count these words in the computation of the performance in Table 7.2. Consequently, the HR3, recall, and precision in the table are actually underestimated while the CL is overestimated for the proposed system.

7.6.4   Results of Single Word Text-to-Image Querying

The single word text-to-image querying results on a set of 500 randomly selected query words are shown in Figure 7.5. The average SWQP (2, 5, 10, 15, 20) values of the proposed system and those of MBRM are reported. A returned image is considered as relevant to the single word query if this word is contained in the ground truth annotation of the image. It is shown that the performance of the proposed probabilistic model has higher overall SWQP than that of MBRM. It is also noticeable that when the scope of the returned images increases, the SWQP(n) in the proposed system attenuates more gracefully than that in MBRM, which is another advantage of the proposed model.

7.6.5   Results of Image-to-Image Querying

From Equations 7.18 and 7.21, it is clear that if we have an image query q_f, based on Equation 7.18, we immediately generate the top m annotation words based on the probability $P(w^j|q_f)$. For each of the m annotation words,

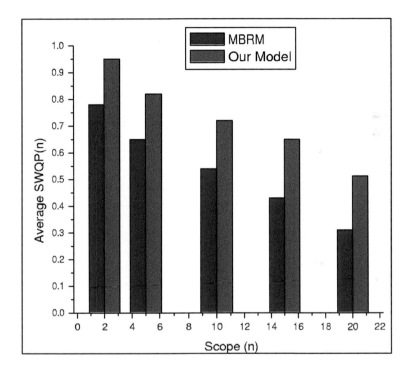

FIGURE 7.5: Average SWQP(n) comparisons between MBRM and the proposed approach. Reprint from [246] ©2006 Springer-Verlag Press and from [245] ©2005 IEEE Computer Society Press.

based on Equation 7.21, we immediately generate an image list based on the probability $P(f_i|w^j)$. Finally, we merge the m ranked lists as the final retrieval result based on the posterior probability $P(f_i|w^j)P(w^j|q_f)$. Clearly, for a general query consisting of words and images, each component of the query may be individually processed and the final retrieval may be obtained by merging all the retrieved lists together based on the posterior probability. For reference purpose, we call this general indexing and retrieval method UP-MIR, standing for *Unified Posterior based Multimedia Information Retrieval*. For the image-to-text annotation and single word text-to-image retrieval, we have reported the evaluations of UPMIR against MBRM. Since the originally reported MBRM method did not include the image-to-image scenario, we report the image-to-image evaluations for UPMIR against UFM [47]. Figures 7.6 and 7.7 document the averaged precision and recall as the performance comparison with UFM using the image-to-image querying mode for the 600 query images on the same evaluation dataset. It is clear that with the pure image querying mode, UPMIR performs at least the same as UFM and in most cases better than UFM (e.g., with the top 2 images retrieved, UPMIR has a 10% better retrieval precision than UFM). To further demonstrate that UPMIR is also more efficient than UFM in image retrieval, we note that UP-MIR and UFM are both implemented and evaluated in the same environment of a Pentium IV 2.26 GHz CPU with 1 GB memory. Given the scale of 17,000 images and 7,736 word vocabulary, the average response time for each query for UPMIR is 0.936 second, while that for UFM is 9.14 seconds. Clearly, UP-MIR beats UFM substantially. This is due to the fact that UPMIR has much lower complexity than UFM, as UFM is region-based and for each comparison between two images UFM requires a combinatorial complexity, while for UPMIR it is only a constant complexity.

7.6.6   Results of Performance Comparisons with Pure Text Indexing Methods

Since the UPMIR performance is biased toward the text component querying, we intend to experimentally justify and demonstrate that UPMIR still offers a better image retrieval than a pure text indexing scheme. For this purpose, we manually evaluate UPMIR using the pure text querying mode (i.e., single word text-to-image querying mode) against Google and Yahoo!. We randomly select 20 words out of the 7,736 word vocabulary and use each of them as a pure text query to pose to UPMIR, Google image search, and Yahoo! image search, respectively. We manually examine the precisions. Figure 7.8 clearly demonstrates that UPMIR outperforms Google and Yahoo! for different numbers of the top images retrieved. Since we do not have access to Google or Yahoo! image databases, though the comparing databases are different in size and content, this is the best we can do to compare their performances. The purpose is to show that a multimodal image mining and retrieval system such as UPMIR still has clear advantages for image retrieval

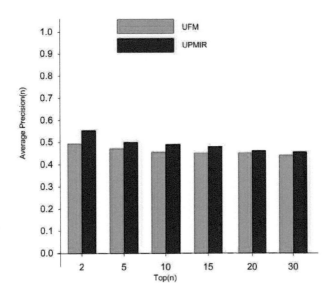

FIGURE 7.6: Precision comparison between UPMIR and UFM.

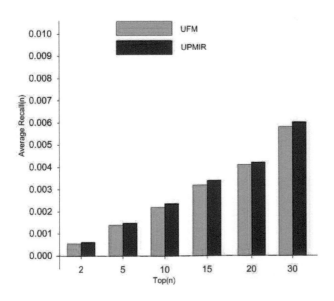

FIGURE 7.7: Recall comparison between UPMIR and UFM.

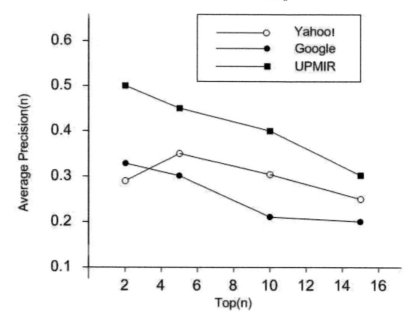

FIGURE 7.8: Average precision comparison among UPMIR, Google Image Search, and Yahoo! Image Search.

over a pure text based indexing system.

7.7   Summary

In this chapter, we have developed a probabilistic semantic model for automatic image annotation and multimodal image data mining and retrieval. Instead of assuming artificial distributions of the annotation words and the unreliable association evidence used in many existing approaches, we assume a hidden concept layer to generate and connect visual features and annotation words. The hidden concept variables are discovered and corresponding probabilities are determined by fitting the generative model to the training set. The model fitting is performed in the criterion of MLE and an EM based iterative learning procedure is developed to approach a local maximum. Based on the model obtained, the image-to-text and text-to-image queryings are conducted in a Bayesian framework, which is adaptive to the dataset and has a clear interpretation of the confidence measure. The proposed model is shown to be promising for image annotation and multimodal image data mining and retrieval; this is demonstrated by the evaluations of the prototype system on

17,000 images and 7,736 textual annotation words automatically extracted from the crawled Web pages. In comparison with a state-of-the-art image annotation system, MBRM, and a state-of-the-art image retrieval system, UFM, it shows the higher reliability and the superior effectiveness of the proposed model and the querying framework, given the noisy and diverse semantics and visual variations of the data we have used.

Chapter 8

Concept Discovery and Mining in a Video Database

8.1 Introduction

In the last two chapters, we have shown examples of concept discovery and mining in an image database in which we use generative models to discover the concepts in the database. In this chapter, we switch the focus to a video database; in addition, we use a combination of the generative models and the discriminative models to discover the concepts in a video database. In particular, we focus on the natural database querying problem in the context of video — the Web video search — and show how concept discovery and mining can help in delivering an effective solution to the Web video search.

Building a video search engine on the Web is a very challenging problem. Compared with the Web page search, video search faces unique challenges (such as a high volume of data for each video and the existence of multimodal information including meta data, visual content, audio, and closed caption). In this chapter, we investigate several promising approaches to boosting the search relevance of a large-scale video search engine on the Web. Specifically, we describe a developed, specialized video categorization framework which combines multiple classifiers based on different modalities; by learning users' querying histories and clicking logs, we propose an automatic query profile generation technique and apply this profile to query categorization; based on this scheme, a highly scalable prototype system is developed, which integrates the online query categorization and offline video categorization. The naive Bayes with a mixture of multinomials, the maximum entropy, and the support vector machine categorization methods and the profile learning technique are evaluated on a large-scale set of video data on the Web. The evaluation of the developed system and user study has indicated that the joint categorization of queries and video data boosts the video search relevance and user search experience. The high efficiency of the proposed approaches is also demonstrated by the good responsiveness of the prototype system for the video search engine on the Web.

The rest of the chapter is organized as follows. Section 8.2 introduces the background and motivation of this research on video search based concept discovery and mining in the video data from the Web. Section 8.3 reviews the related work from the literature. In Section 8.4 the video categorization framework based on the meta data and content features is described. Three classifiers we have developed are introduced in this section. Section 8.5 presents the query profile generation technique and the application of the technique to query categorization. The implementation issues of the system, the characteristics of the training and testing data set, the experiments to evaluate the proposed approach, and the comparison study of the classifiers, both on recall/precision and on computation complexity, are reported in Section 8.6. Finally, this chapter is concluded in Section 8.7.

8.2   Background

The task of a video search engine on the Web is to search a large amount of video clips on the Web for users' queries. To develop a successful video search engine, several factors are essential. First, the coverage of the searchable video clips should be extremely large such that most video clips on the Web can be queried. Second, the relevance of the search results should be sufficiently high to be useful and to be personalized to individual users. Third, the search system must be highly scalable, and the response time of users' queries should not be dependent upon (at least not linearly dependent upon) the size of the video set. Compared with already successful Web page search technologies, video search technology on the Web is still in its infant stage. Although video search and Web page search share many basic characteristics of information retrieval and data mining, there are several unique difficulties of video search (as well as multimedia search).

One difficulty is that the semantics of a video on the Web is typically not explicitly labeled such that it is very difficult to be precisely indexed. On the other hand, video is much more content-rich than other media in that it contains additional information such as visual, audio, and closed caption. Though content-based video (multimedia) retrieval has been under intensive research for more than a decade [98] and a large number of features [99, 190, 185] and similarity metrics have been proposed, the success is rather limited due to the notorious semantic gap [192]. On the other hand, for video data mining and retrieval, the query-by-text and recently the query-by-concept [10] paradigms have proven very powerful and convenient for users to search the content, while the query-by-example paradigm used has not shown clear attractiveness to general users for the video search scenario. To reduce the semantic gap, multimedia data mining using multimodal information (e.g., text, audio, video, and image) has been proposed and shown to be very promising [245]. The objective of this research is to develop a practically usable video search engine on the Web which can meet the above requirements.

By studying the query logs of a video search engine, we have found that most Web users search for video clips falling in certain categories (e.g., News, Movie, Music) but they typically just input very short query words (more than 90% of the queries contain fewer than three words). For example, users searching for "hurricane katrina" actually look for news video clips about the recent Hurricane Katrina instead of education videos about the generation of hurricanes instructed by Katrina; users searching for "Madonna" are most likely interested in music video clips of the pop star Madonna, instead of some funny videos of a person whose name is Madonna. The motivation of this study is to infer the categories of users' queries and to use these categories further to guide or constrain the video search. In addition, the video clips to be searched are automatically categorized. In this way, the returned video clips fit users' querying needs better. This can be done by utilizing the well-labeled training video data and users' querying histories to boost the search relevance. This chapter proposes and evaluates an approach to the problem by jointly categorizing video clips and users' queries automatically.

Specifically, in this chapter,

1. we describe a specialized video categorization framework which combines multiple classifiers based on both meta data and content features; three categorization learning methods are developed and evaluated, including the naive Bayes classifier with a mixture of multinomials [173], the maximum entropy classifier [159], and the support vector machine [207] classifier;

2. by learning users' querying histories and clicking logs, we propose an automatic query profile generation technique and apply the profile to query categorization; and finally,

3. a highly scalable prototype system is developed, which integrates the online query categorization and offline video categorization; different classifiers and the profile learning technique are compared and the performance differences are studied; and the prototype system is extensively evaluated on a large-scale set of hundreds of queries and millions of video clips on the Web and shows significant improvement on search accuracy.

8.3 Related Work

The video classification concerned in this chapter is in the context of a video search engine. It belongs to a *wide input domain* video classification. A wide domain has an unlimited and unpredictable variability in its appearance even for the same semantic meaning [192]; such domains include broadcasting news video, sports video, mainstream entertainment video, etc. This video classification problem has been under intensive research [61, 204] since the mid-1990s. One of the first attempts was the work by Fischer et al [78]. They present a three-step approach to classify news, commercial, cartoon, and sports video clips. The first step is collecting basic acoustic and visual statistics; the second step attempts to derive style attributes (SA) which include scene length, camera motion intensity, caption detection, etc; finally, the distributions of these SAs are used in the discrimination of the video genres. While the approach is simple, the classification is based on ad hoc rules; thus, the system is not robust and cannot be generalized well. Most of the existing video classification work focuses on the low-level feature representation of the video [112, 111], and little work has integrated the video classification into a video search engine and has investigated the impacts to the search relevance.

How to provide more context to user queries to develop a personalized search experience is another hot topic in the search engine research. Many researchers address this problem from the information filtering [218, 39] and intelligent agent [165] perspectives. Most of them recommend documents using the user profiles [6]. However, the technique we investigate is to retrieve categories of interest for a user query and to guide the user to the category of video clips, matching the user's querying interests. Although capturing users' querying needs for multimedia content is much more complicated than for the general Web search, user research shows that text based queries still are more accurate and convenient than the queries based on other formats (e.g., queries based on examples, queries based on sketches) for users to search multimedia content on the Web. No related work is reported yet on users' query classification without users' explicit awareness for multimedia search applications on the Web.

Considering that video clips have rich content cues to explore for classification and retrieval, many aggregation approaches have been developed to combine the search results from multiple modalities. Lin and Hauptmann [139] have proposed a meta-classification combination strategy. An SVM based meta-classifier is constructed by concatenating the outputs from an ensemble of unimodal classifiers. Comparing the probability-based approaches [127], this approach delivers superior results. More recently, a query-class dependent weighting scheme has been proposed for the combination of video retrieval results [224] from multiple modalities. In this approach, different weights are learned from users' queries for different modalities by modeling the problem as a maximum likelihood estimation problem and using the Expectation-Maximization (EM) technique to solve the problem. The intuition is simple, but how good the EM learning fits the shots to the hidden variables is not clear, and the performance is sensitive to the accuracy of query classification. What is more important is that the above two approaches are evaluated on only a few hundred video clips. Due to their high computational complexity, they are not able to scale up to handle millions of queries every day on tens of millions of video clips that a video search engine for the Web faces.

Wu et al have proposed a two-step optimal multimodal fusion approach in the context of the video classification task [222, 223]. In this approach, first a set of statistically independent modalities is found from the raw features extracted from multiple media sources by using PCA, ICA, and independent modality grouping (IMG), consecutively. Then a cascade of SVM classifiers is used to determine the optimal combination of the individual modalities, which is called super-kernel fusion. The non-linear fusion is achieved by exploiting the kernel nature of SVM and an improved classification accuracy is reported in the experiments. Although in this approach the interdependency among features is exploited and the non-linear fusion is employed, the factors that affect the fusion performance substantially (i.e., modality independence, curse of dimensionality, and fusion-model complexity) are manually tuned.

Since 2001, the "TRECVID" [108] has been a benchmark for evaluating video retrieval systems. The search topics in "TRECVID" contain not only text but also possible examples (including video, audio, and images) which represent the user's information need. We note that some benchmark metrics in "TRECVID" are not relevant to the video search on the Web because the quality and content of video clips on the Web are much more heterogeneous than the video corpus of "TRECVID"; in addition, the Web search querying form is different from the querying forms used in "TRECVID". For example, the text query form to search video clips on the Web is intuitive and powerful for users, while query-be-example has not been widely accepted for video search engines on the Web.

The objective of the work described in this chapter is to develop a framework of joint categorization of queries and video clips for Web-based video search by mining the multi-modalities of video contents on the Web for concept discovery. The very large scale of Web-based video search and the high efficiency constraint of joint query and video categorization have not been seriously addressed in the literature before.

8.4 Video Categorization

To classify videos, we use multimodal features (i.e., text features and video content features) and apply multiple classification models for evaluation and comparison. Collecting the meta data and extracting the video content features from the training video data, we train two classifiers, one for each modality, respectively. Different classification models are applied and modifications on these models are conducted to adapt to the specific task of video search. The system architecture of the framework of the joint categorization of queries and video clips we have proposed is shown in Figure 8.1.

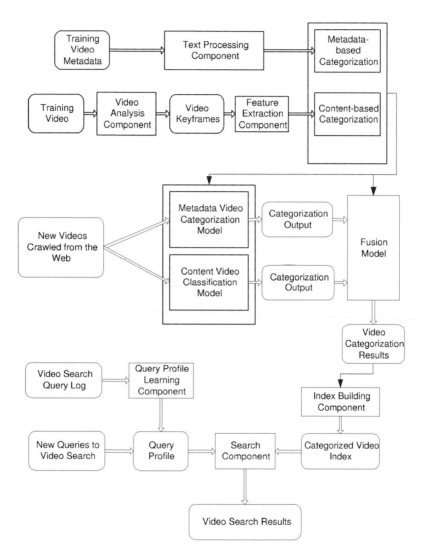

FIGURE 8.1: The architecture of the framework of the joint categorization of queries and video clips. Reprint from [239] ©2006 ACM Press.

8.4.1 Naive Bayes Classifier

Naive Bayes [66] is a well-studied classification technique and is studied in Section 3.2.4. As discussed in Section 3.2.4, despite the strong independence assumptions, its attractiveness comes from the low computational cost, relatively low memory consumption, and the ability to handle heterogeneous features and multiple categories. In the video categorization with collateral textual data, the distribution of the words for each textual field of the video's meta data is modeled as a multinomial distribution. A textual field is treated as a sequence of words, and it is assumed that every word position is generated independently of each other. Consequently, each category has a fixed set of multinomial parameters. The parameter vector for a category c is $\vec{\theta}_c = \{\theta_{c1}, \theta_{c2}, \ldots, \theta_{cn}\}$, where n is the size of the vocabulary, and θ_{ci} is the probability that word i occurs in that category with the constraint $\sum_i \theta_{ci} = 1$. The likelihood of a video passage is a product of the parameters of the words that appear in the passage:

$$p(o|\vec{\theta}_c) = \frac{(\sum_i \sum_k w_k t_{i,k})!}{\prod_{i,k}(w_i t_{i,k})!} \prod_{i,k}(\theta_{ci})^{w_k t_{i,k}} \tag{8.1}$$

where $t_{i,k}$ is the frequency count of word i in the field k, whose weight is w_k, of video object o. Here we take into consideration of the field importance weight w_k because it is observed from the video clips on the Web that different fields of the video meta data have different contributions to describing the semantics of the video clips on the aspects of precision and discrimination capability. This adjustment of the model improves the video categorization accuracy, as demonstrated in the experiments reported in Section 8.6. By assigning a prior distribution over the set of classes, $p(\vec{\theta}_c)$, we arrive at the minimum-error categorization rule [66] which selects the category with the largest posterior probability,

$$l(o) = \arg\max_c [\log p(\vec{\theta}_c) + \sum_i \sum_k w_k t_{i,k} \log \theta_{ci}]$$

$$= \arg\max_c [b_c + \sum_i \sum_k w_k t_{i,k} z_{ci}] \tag{8.2}$$

where b_c is the threshold term and z_{ci} is the category c weight for word i. These values are natural parameters for the decision boundary. The parameters $\vec{\theta}_c$ are estimated from the training data. This is done in the prototype system by selecting a Dirichlet prior and taking the expectation of the parameter with respect to the posterior. This gives us a simple form for the estimate of the multinomial parameter, which involves the field-weighted number of times word i appearing in the passages of the video clips belonging to class c ($\sum_k w_k N_{i,k,c}$, where $N_{i,k,c}$ is the number of the times word i appears in the field k of the video clips in category c), divided by the total field-weighted number of word occurrences in field k of class c ($\sum_k w_k N_{k,c}$). For word i, a prior adds in α_i imagined occurrences so that the estimate is a smoothed version of the maximum likelihood estimate:

$$\vec{\theta}_{ci} = \frac{\sum_k w_k N_{i,k,c} + \alpha_i}{\sum_k w_k N_{k,c} + \alpha} \tag{8.3}$$

where α denotes the sum of α_i. While α_i can be set differently for each word, we follow the common practice by setting $\alpha_i = 1$ for all words.

In video classification based on the visual content, each feature dimension v_d is modeled as a Gaussian in category c:

$$p(v_d|c) = \frac{1}{\sqrt{2\pi}\sigma_{c,d}} \exp[-\frac{(v_d - m_{c,d})^2}{2\sigma_{c,d}^2}] \tag{8.4}$$

where $m_{c,d}$ is the mean value of v_d, and $\sigma_{c,d}$ is the standard deviation of v_d in category c, respectively. Applying the maximum-likelihood method [25] on the training video data for each category c, we obtain the following unbiased estimations of the mean $m_{c,d}$ and the standard deviation $\sigma_{c,d}$.

$$\widehat{m}_{c,d} = \frac{1}{U_c} \sum_{i \in c} v_{i,d} \tag{8.5}$$

and

$$\widehat{\sigma}_{c,d}^2 = \frac{1}{U_c - 1} \sum_{i \in c} (v_{i,d} - \widehat{m}_{c,d})^2 \tag{8.6}$$

where $v_{i,d}$ denotes the d-th dimension of the feature vector v_i and U_c is the number of the video clips belonging to category c. Given the assumption that the visual features are conditionally independent for category c, the categorization is performed based on a similar formula to Equation 8.2.

8.4.2 Maximum Entropy Classifier

Maximum entropy [116] is a general technique for estimating a probability distribution from data. The overriding principle in maximum entropy is that when nothing is known, the distribution should be as uniform as possible, that is, should have maximal entropy. The maximum entropy classifier [45] estimates the conditional distribution of the category label given a video with constraints specified in the training data. Each constraint expresses a characteristic of the training data that should also be present in the learned distribution. In a generalized form, each video object o in a category c is represented by $\vec{f}(o,c) = \{f_1(o,c), f_2(o,c), \ldots, f_n(o,c)\}$. Maximum entropy allows us to restrict the model distribution to have the same expected value for feature $f_i(o,c)$ as seen in the training data. Thus, we stipulate that the learned conditional distribution $p(c|o)$ must have the property:

$$\frac{1}{U} \sum_o f_i(o, c(o)) = \sum_o p(o) \sum_c p(c|o) f_i(o, c) \tag{8.7}$$

where U is the number of the training video clips. Notice that the video distribution $p(o)$ is unknown, and we are not interested in modeling it. Thus, we use the training data, without category labels, as an approximation to the video distribution, and enforce the constraint as:

$$\frac{1}{U} \sum_o f_i(o, c(o)) = \frac{1}{U} \sum_o \sum_c p(c|o) f_i(o, c) \qquad (8.8)$$

Here the feature $f_i(o, c)$ is either the normalized word counts for the meta data or the visual feature we have extracted from the video frames. For each feature, we measure its expected value over the training data and take this as a constraint for the model distribution.

When constraints are estimated in this fashion, it is guaranteed that a unique distribution exists that has the maximum entropy. Moreover, it can be shown [168] that the distribution is always of the exponential form:

$$p(c|o) = \frac{1}{Z(o)} \exp(\sum_i \lambda_i f_i(o, c)) \qquad (8.9)$$

where λ_i is a parameter to be estimated and $Z(o)$ is simply the normalizing factor to ensure a proper probability:

$$Z(o) = \sum_c \exp(\sum_i \lambda_i f_i(o, c)) \qquad (8.10)$$

The form of the maximum entropy classifier is a multi-category generalized form of the logistic regression classifier [151]. When the constraints are estimated from the labeled training data, the solution to the maximum entropy problem is also the solution to a dual maximum likelihood problem for models of the same exponential form. The attractiveness of this model is that it is guaranteed that the likelihood surface is a convex manifold with a single global maximum and no local maxima. We perform a hill-climbing algorithm [168] in the likelihood space to find the global maximum. To reduce the overfitting, we introduce a Gaussian prior on the model, with zero mean, and a diagonal covariance matrix. This prior favors feature weightings that are closer to zero, that is, less extreme. The prior probability of the model is just the product over the Gaussians of all the feature values λ_i with variance σ_i^2:

$$p(\Lambda) = \prod_i \frac{1}{\sqrt{2\pi\sigma^2}} \exp(\frac{-\lambda_i^2}{2\sigma_i^2}) \qquad (8.11)$$

It is shown [45] that introducing a Gaussian prior to each λ_i improves performance for the language modeling tasks when sparse data causes overfitting. Similar improvements are also demonstrated in the reported experiments in Section 8.6.

8.4.3   Support Vector Machine Classifier

Unlike the above generative models, as studied in Section 3.7, the classic support vector machine (SVM) [207] is a binary categorization method based on a discriminative model which implements the *structural risk minimization* (SRM) principle. It creates a classifier with minimized *Vapnik-Chervonenkis* (VC) dimension. SVM minimizes an upper bound on the generalization error rate. The attractiveness of SVM comes from its good generalization performance for pattern classification problems without needing to incorporate the domain knowledge. We formulate the video categorization problem as an ensemble of binary categorization problems, with one SVM classifier for each category. For a binary categorization problem, if the two categories are linearly separable, the hyperplane that does the separation can be easily calculated by $\vec{w}^T o + b = 0$, where $\vec{w}$ is a weight vector and b is a bias. The goal of SVM is to find the parameters $\vec{w}$ and b for the optimal hyperplane to maximize the distance between the hyperplane and the closest data point (i.e., the support vectors):

$$(\vec{w}^T o + b)c \geq 1 \tag{8.12}$$

If two categories are non-linearly separable, the input vectors may be non-linearly mapped to a higher-dimensional *feature space* by an inner-product kernel function $K(\vec{x}, \vec{x}_i)$. Here the *feature space* is a convention name in SVM literature, which is different from the features we commonly use to represent the video clips. Typical kernel functions are polynomial, radial-basis, and sigmoid [207]. An optimal hyperplane is constructed for separating the data in the higher-dimensional *feature space*. The hyperplane is optimal in the sense of being a maximal margin classifier with respect to the training data (see Section 3.7).

In its standard formulation, SVM only outputs a prediction +1 or -1, without any associated measure of confidence. In the proposed prototype system, we consider a special modification of SVM, which can output a posteriori category probability. This modification retains the powerful generalization capability of SVM and paves the way to wide extensions, such as an integration within a probabilistic framework. Probabilistic extensions of the SVM, where an associated probability of category membership is output, have been independently suggested in the literature. For this work, we use a probabilistic version of the SVM (PSVM) similar to the one proposed in [230]. Here, the probability of the membership in category $y, y \in \{+1, -1\}$ is given by:

$$p(y|o) = \frac{1}{1 + \exp(-yA(\vec{w}^T o + b))} \tag{8.13}$$

where A is the parameter to determine the slope of the sigmoid function. This modified SVM retains exactly the same decision boundary as defined by $\vec{w}^T o + b = 0$, yet allows an easy computation of the posterior category probabilities. The output of PSVM can be compared with the output of other generative model based categorization methods. We use a cross-validation scheme to set this parameter A for each category. In the proposed prototype system, we apply a PSVM classifier on both the meta data and the content feature of the training video data for each category.

8.4.4   Combination of Meta Data and Content Based Classifiers

After constructing classifiers based on the meta data and the content features of the video clips, ideally the classifiers of the two modalities are complementary to each other and we intend to combine the categorization outputs from the two modalities to boost the accuracy. Therefore, the problem of selecting the most effective classifiers and determining the optimal combination weights naturally follows. The experiments in Section 8.6 show that for some categories (e.g., news video, music video), the meta data based classifiers have a better accuracy than the content feature based classifiers, while for other categories (e.g., adult video), the content feature based classifiers work better. To take advantage of this prior knowledge, we propose a voting-based category-dependent combination scheme to develop a fused output. Specifically, each video clip can have multiple labels (e.g., a financial news video belongs to both the news video category and the finance video category). Hence, we build a binary classifier for each category and in the training phase we perform a *k-fold validation* procedure to obtain an estimated categorization accuracy $a_{i,m}$ for each category c_i by the classifier based on modality m. The combination scheme is:

$$p(c_i|o) = \frac{\sum_m a_{i,m} p_m(c_i|o)}{\sum_m a_{i,m}} \qquad (8.14)$$

The video is assigned to category c_i if $p(c_i|o)$ is larger than a threshold; $a_{i,m}$ reflects the effectiveness of the modality m to the category c_i, while $p_m(c_i|o)$ is the confidence of assigning o to category c_i by the modality m based classifier. This scheme is a validation accuracy weighted combination scheme, and the strengths of the classifiers based on both modalities are integrated, which improves the performance of the final categorization recall and precision. This proposed fusion approach is easy to implement and is computation-efficient. In contrast, the highly expensive computation complexity of the approach in [224] makes it not practical for a search engine with the scale of several dozens of queries per second and tens of millions of video clips indexed, while the gain of the recall-precision using the approach is rather limited in the experiments [224]. Compared with the approach in [222], this fusion approach safely removes the expensive procedures of PCA, ICA, and IMG because the features we use are statistically independent by applying a mutual information based feature selection [133]. Another important advantage of this fusion model is its lower complexity. The super-kernel fusion in [222] has much higher complexity and, thus, may jeopardize the model's generalization capability.

In addition, the expensive computation cost precludes its application to real-time classification in practice.

8.5   Query Categorization

How to personalize users' video search experiences is a huge challenge. When the same query is submitted by different users, a typical search engine returns the same result, regardless of who submitted the query. This may not be appropriate for users with different information needs. For example, for the query "apple", some users may be interested in video clips dealing with "apple gardening" as "fruit gardening", while other users may expect news or financial video clips related to Apple Computers. One way to disambiguate the words in a query is to manually associate a small set of categories with the query. However, users are often too impatient to identify the proper categories before submitting queries. We propose an approach to supplying a small set of categories as a context for each query submitted by a user, based on the users' querying histories. Specifically, we provide a strategy to model and gather users' search histories and construct a query profile. Based on the query profile, appropriate categories are automatically deduced for each user query.

To construct the query profile, we analyze the query logs of users in the search engine. The histories of users' queries and their corresponding clicked video results are extracted from the log.[1]. From the log we generate two matrices, VT and VC, as shown in Table 8.1.

Each cell in VT denotes the significance of the term in the description of relevant video clips clicked by users, which is computed by the standard information retrieval techniques (*TF-IDF*) [12]. VC is generated by Web surfers to describe the relationships between the categories and video clips. What we intend to generate is the query profile matrix QP as shown in Table 8.2.

To learn QP from VT and VC, we apply a method based on *linear least square fitting* (LLSF) [226], in which QP is computed such that $VT \times QP^T \approx VC$ with the least sum of square errors. Solving the problem by employing Singular Value Decomposition (SVD) [56], we obtain:

$$QP = VC^T \times U \times S^{-1} \times V^T \tag{8.15}$$

where the SVD of VT is $VT = U \times S \times V^T$; U and V are orthogonal matrices and S is a diagonal matrix.

---

[1]We also analyze the query log of the Web search to construct the query profile because users' semantic querying needs are represented similarly for any vertical search.

Table 8.1: Matrix representation of users' query log. Reprint from [239] ©2006 ACM Press.

(a) Matrix VT. V1 to V4 are video clips.

Video \ Term	Tom Cruise	movie	Hollywood	football	Super Bowl	touch down
V1	1	1	0.8	0	0	0
V2	0.3	0.8	0.6	0	0	0
V3	0	0	0	1	0	1
V4	0	0	0	0.62	0.7	0.3

(b) Matrix VC. V1 to V4 are video clips.

Video \ Category	Movie	Sport
V1	1	0
V2	1	0
V3	0	1
V4	0	1

Table 8.2: Matrix representation of query profile QP. Reprint from [239] ©2006 ACM Press.

Video \ Term	Tom Cruise	movie	Hollywood	football	Super Bowl	touch down
Movie	0.7	1	0.9	0	0	0
Sport	0	0	0	1	0.67	0.55

For each query term, we predict its related categories by using QP and categorize it accordingly. Specifically, the similarity between a query vector q and each category vector qp in the query profile QP is computed by the *Cosine* function [12]. Then the categories are ranked in descending order of the similarities and the top-ranked queries are provided to the users for selecting the one as their query's context.

8.6   Experiments

To evaluate the joint categorization of videos and queries and its impact on the search relevance, we have built a testbed.

8.6.1   Data Sets

The objective is to develop a video search engine on the Web. Video clips on the Web have different characteristics in comparison with TREC video data. First, typically the video clips on the Web have considerable but diverse meta data and are highly heterogeneous w.r.t. the visual content and semantics, but the duration of the video clips is typically short. On the other hand, TREC video data do not have much meta data and most of the TREC video data are broadcasting news videos [98]. In this sense, the TREC video data cannot represent well the video data on the Web. Second, the number of shots in the TREC video corpus is far smaller (e.g., only 32,318 video shots in the reference set of TREC-2003) than the number of the video clips on the Web (i.e., hundreds of millions scale). For these reasons, instead of using the TREC video test corpus, we have crawled the video clips from the Web to compose the test corpus. The number of the video clips in our test corpus is much larger than that of the TREC video data. Using this test video set, we have evaluated the prototype video search engine in a real Web search environment.

In the testbed, we have collected more than 400,000 video clips and have associated the meta data from the Web. Among them, 385,739 pre-classified video clips belong to five categories (News, Music, Movie, Finance, and Funny Video)[2]. These video clips compose the training set, which has 2000 GB AVI files and a total duration of 15,572 hours (2.34 minutes/video on average). For text features, we have collected the surrounding text in the Web pages containing the video clips on the Web along with the accompanying meta data (e.g., file names, titles, closed captions, and user tags) which annotate the video clips. The text features of each video clip are composed of a number

---

[2]In the training set, each video is labeled to one and only one category

of fields, while each field has an associated weight and is either a text passage or a categorical field. Due to the complex nature of the video content, further investigation is necessary on content feature representation (see Chapter 2 for a systematic study on the multimedia features). At this time, many content features (e.g., optical flow, image sequence, video OCR, and audio) are reasonably meaningful for the classification/search applications. Since the objective of this research is not for feature selection, we use several easy-to-compute yet effective content features extracted from the video data. To train content-based classifiers, all the video clips in the testbed are segmented and one most representative keyframe is extracted for each video clip [68]. We employ this representation of video clips due to its simplicity and low computation cost. Although simple, this representation proves to be effective in the experiments. To represent the spatial color distribution of the keyframes in the video data, color auto-correlograms [112] are computed. As studied in Section 2.3.1.3, color auto-correlograms compute a histogram of color pairs in different distances. Formally, it is defined as

$$\Gamma_{c_i,c_i}^{(k)}(I) \triangleq |\{p_1 \in I_{c_i}, p_2 \in I_{c_i} | \|p_1 - p_2\| = k\}| \tag{8.16}$$

where $|p1 - p2|$ is the *L1* distance between pixel p_1 and p_2 whose color is in the bucket c_i. It has been shown that the auto-correlograms are effective in image retrieval [118]. Another content feature we have extracted is the texture feature for a keyframe. To represent the texture feature, as studied in Section 2.3.1.4, we uniformly partition each keyframe into blocks and compute Gabor wavelet coefficients [145] by a filter bank for each block. A two-dimensional Gabor function $g(x, y)$ and its Fourier transform can be written as:

$$g(x,y) = (\frac{1}{2\pi\sigma_x\sigma_y}) \exp[-\frac{1}{2}(\frac{x^2}{\sigma_x^2} + \frac{y^2}{\sigma_y^2}) + 2\pi jWx]$$

$$G(u,v) = \exp\{-\frac{1}{2}[\frac{(u-W)^2}{\sigma_u^2} + \frac{v^2}{\sigma_v^2}]\}$$

where $\sigma_u = 1/2\pi\sigma_x$, $\sigma_v = 1/2\pi\sigma_y$, and W denotes the upper center frequency of interest. Based on the mother Gabor wavelet $g(x, y)$, a self-similar filter dictionary may be obtained using appropriate dilations and rotations of $g(x, y)$ through the generating function:

$$g_{mn}(x,y) = a^{-m}G(x',y'), \, a > 1, \, m, n = \text{integer}$$

$$x' = a^{-m}(x\cos\theta + y\sin\theta), \text{ and } y' = a^{-m}(-x\sin\theta + y\cos\theta)$$

where $\theta = n\pi/K$ and K is the total number of orientations. The scalar factor a^{-m} is meant to measure the energy that is independent of m, $m = 0, 1, \ldots, S-1$. By using the filter response for S scalars and K orientations, we obtain a vector for each block which describes the texture feature. The color auto-correlograms and Gabor wavelet coefficients are combined to compose the content features for the keyframe of a video clip.

In addition to the above video clips database, we have also collected a video database in which 10,000 video clips are tagged as offensive video clips. Among them, 7,000 video clips are used for training. For the meta data based classification, standard text processing is performed, including upper-lower case conversion, stop word removal, phrase detection, and stemming.

8.6.2   Video Categorization Results

First, we investigate the effectiveness of the three categorization models based on the meta data and the content features, respectively. In the proposed prototype system, to accommodate that each video may have multiple labels, we train a binary classifier for each category, and the final labels of a video clip are those categories whose binary classifiers return positive. For each experiment, the 10-fold cross-validation strategy [153] is employed to validate the models learned. For the meta data based categorization, the number of features is very large (58,732 in our experiments); to improve the time/space performance and to reduce the over-fitting problem, we have performed the feature selection based on the mutual information [133] and, consequently, the optimal number of the features is determined by the cross-validation. For the content feature based categorization, on the other hand, we do not perform the feature selection because the dimensionality of the features is not high (192 dimensions). The confusion matrices for the validation data of the meta data based classifiers are listed in Tables 8.3 to 8.5. The variance of the Gaussian prior in the maximum entropy classifiers is empirically set and the iterations are controlled by an upper bound of the relative increase in the log-likelihood and an upper bound of the number of the iterations. Since we need to train a large number of video clips and the categorization must be fast, the linear SVM classifier described in [124] is used due to its high efficiency. From the experimental results, the maximum entropy classifier gives better recall (4.8% margin) and precision (6.47% margin) than the naive Bayes classifier. The reason could be partly due to the too-strong assumption made in naive Bayes that each feature is generated independently of each other. Moreover, the exponential form of the posterior probability in the maximum entropy classifiers fit the data better and higher log-likelihoods are obtained than in the naive Bayes classifiers. The best performance is achieved by the SVM classifiers. This demonstrates the advantage of the discriminative models over the generative models for the categorization task in this scenario. From the tables, we see that a significantly higher number of video clips which have one true label are categorized to multiple labels (e.g., news to news and music; movie to movie and music; and movie to movie, music, and funny) using the naive Bayes or the maximum entropy classifiers than using the SVM classifiers. This shows that the discrimination capability of SVM is better than its two peers. The difference can be explained by the

Table 8.3: Confusion matrix of the naive Bayes classifier based on the meta data. The confusion matrix is a modified version of the classic confusion matrix to accommodate the multi-label categorization results. "A" denotes actual category and "P" denotes predicted category. "Negative" denotes that no classifiers give position predictions. Legend: a – news video, b – finance video, c – movie, d – music video, e – funny video, f – negative video, t – total, p – precision, r – recall, and x – not applicable. Reprint from [239] ©2006 ACM Press.

P\A	a	b	c	d	e	t	p
f	35	0	6	11	0	52	0
a	25480	1	8	2	0	25491	99
a, b	185	27	0	0	0	212	0
a, d	126	0	0	19	0	145	0
a, e	13	0	0	0	0	13	0
b	57	19460	2	0	0	19519	99
c	32	0	23345	3	0	23380	99
c, a	717	0	221	0	0	938	0
c, a, d	14	0	0	6	0	20	0
c, d	4	0	1794	3402	0	5200	0
c, d, e	0	0	4	0	0	4	0
c, e	0	0	71	0	1	72	0
d	15	0	0	7768	0	7783	99
e	0	0	0	0	2910	2910	100
t	26678	19488	25451	11211	2911	85739	x
r	95	99	91	69	99	x	x

different optimization objectives in the two different types of the models. SVM optimizes the categorization accuracy explicitly by maximizing the margins between the partition hyperplane and two categories, while both the naive Bayes and the maximum entropy classifiers maximize the log-likelihood of the data set for the models learned. Although the models obtained have the highest probability to generate the data in hand, the categorization errors are not to be minimized for the models. Figure 8.2 shows the comparisons of the average categorization recalls and precisions for the three models.

For the content feature based categorization, we use the same metrics and the results are a little different. Figure 8.3 shows the average categorization accuracies using the different classifiers. The performance of the content feature based categorization is worse than that of the meta data based categorization (but still reasonable), which is what we have expected. The SVM classifier performs the best as in the meta data based categorization. Interestingly, the naive Bayes classifier performs a little better (3.02% margin) than the maximum entropy classifier. We assume that each dimension of the content feature conforms to Gaussian distribution in the naive Bayes learning. This may be explained that the maximum entropy classification is more sensitive

Table 8.4: Confusion matrix of the maximum entropy classifier based on the meta data. The confusion matrix is a modified version of the classic confusion matrix to accommodate the multi-label categorization results. "A" denotes actual category and "P" denotes predicted category. "Negative" denotes that no classifiers give position predictions. Legend: a – news video, b – finance video, c – movie, d – music video, e – funny video, f – negative video, t – total, p – precision, r – recall, and x – not applicable. Reprint from [239] ©2006 ACM Press.

P\A	a	b	c	d	e	t	p
f	181	160	158	175	10	684	0
a	26094	0	42	0	0	26136	99
a, b	123	0	0	0	0	123	0
a, b, d	2	0	0	0	0	2	0
a, c	91	0	0	2	0	93	0
a, e	5	0	0	0	0	5	0
b	24	19328	5	0	0	19357	99
c, b	0	0	1	0	0	1	0
c, d	0	0	25	29	0	54	0
c, e	0	0	2	0	0	2	0
c	108	0	25175	75	0	25358	99
c, a	40	0	9	0	0	49	0
d	10	0	33	10930	0	10973	99
e	0	0	1	0	2901	2902	99
t	26678	19488	25451	11211	2911	85739	x
r	97	99	98	97	99	x	x

Table 8.5: Confusion matrix of the SVM classifier based on the meta data. The confusion matrix is a modified version of the classic confusion matrix to accommodate the multi-label categorization results. "A" denotes actual category and "P" denotes predicted category. "Negative" denotes that no classifiers give position predictions. Legend: a – news video, b – finance video, c – movie, d – music video, e – funny video, f – negative video, t – total, p – precision, r – recall, and x – not applicable. Reprint from [239] ©2006 ACM Press.

P\A	f	a	b	c	d	e	t	r
f	0	1	0	2	6	0	9	0
a, d	0	5	0	0	0	0	5	0
a	0	26669	0	0	0	0	26669	100
b	0	0	19486	0	0	0	19486	100
b, d	0	0	1	0	0	0	1	0
c, b	0	0	1	0	0	0	1	0
c, d	0	0	0	2	0	0	2	0
c	0	0	0	25442	0	0	25442	100
c, a	0	2	0	1	0	0	3	0
c	0	1	0	4	11205	0	11210	99
e	0	0	0	0	0	2911	2911	100
t	0	26678	19488	25451	11211	2911	85739	x
r	x	99	99	99	99	100	x	x

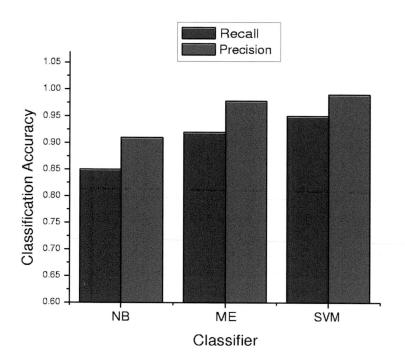

FIGURE 8.2: Comparisons of the average classification accuracies for the three classifiers based on the meta data. (NB: Naive Bayes; ME: Maximum Entropy; SVM: Support Vector Machine.) Reprint from [239] ©2006 ACM Press.

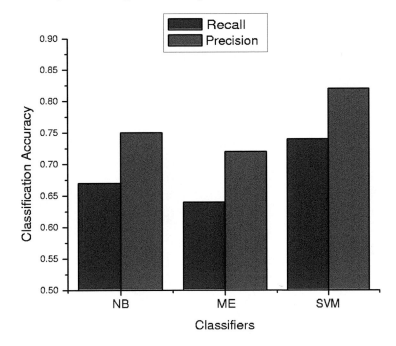

FIGURE 8.3: Comparisons of the average classification accuracies for the three classifiers based on the content features. Reprint from [239] ©2006 ACM Press.

to feature selection than the naive Bayes and that the content features may be more natural for the naive Bayes than for the maximum entropy. Another observation is that for the offensive video category, the content feature based classifiers have a much better categorization accuracy than the meta data based classifiers. Figure 8.4 compares the recalls/precisions of the SVM classifier based on the meta data and the SVM classifier based on the content features for the offensive video category. This observation substantiates our motivation that different modalities of video can complement each other for a better accuracy. By applying the proposed voting-based category-dependent combination scheme, the categorization accuracy is boosted, as we have expected. The results are reported in Figure 8.5.

8.6.3 Query Categorization Results

Apart from the metrics used in "TRECVID", such as the mean average precision (MAP) and the average reciprocal rank (ARR) [98], to measure the query categorization accuracy, the following performance metric is used in the

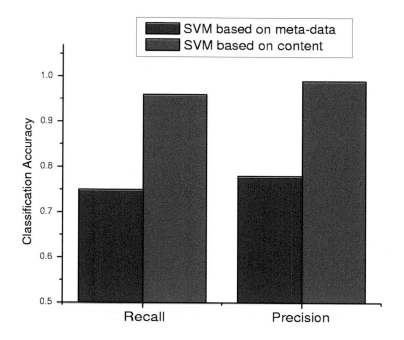

FIGURE 8.4: The comparisons of the categorization accuracies of SVM based on the meta data and the content features for the offensive video category. Reprint from [239] ©2006 ACM Press.

Category		Meta-data	Content Feature	Combine
News	Recall	0.92	0.72	0.92
	Precision	0.90	0.74	0.91
Music	Recall	0.95	0.76	0.96
	Precision	0.98	0.74	0.96
Movie	Recall	0.96	0.72	0.97
	Precision	0.98	0.65	0.98
Offensive Video	Recall	0.82	0.97	0.98
	Precision	0.62	0.99	0.99

FIGURE 8.5: Comparisons of the categorization accuracies for different modalities. Reprint from [239] ©2006 ACM Press.

Table 8.6: Query categorization results. Reprint from [239] ©2006 ACM Press.

(a) Query examples in the test set and their classification results.

#	Query	Returned categories
q1	Michael Jackson	News, Music, Funny
q2	Car race	Movie, Funny, News
q3	World of the war	Movie, News, Finance
q4	Iraq war	News, Finance, Music
q5	Great wall in China	Funny, News, Kids

(b) QC for the query examples and the 100 sampled queries in the test set.

	q1	q2	q3	q4	q5	100 test queries
QC	0.826	0.784	0.923	0.585	0.892	0.845

experiments:

$$QC = (\sum_{c \in topN} s_c)/T = (\sum_{c \in topN} \frac{1}{1 + rank_c - Trank_c})/T \qquad (8.17)$$

where T is the number of the related categories to the query and s_c is the score of a related category c that is ranked among the top N returned categories. We set N as 3 in the experiments. $rank_c$ is the rank of c and $Trank_c$ is the highest possible rank for c. We have randomly sampled 100 queries and have computed QC for each query. For example, assume that $c1$ and $c2$ are related categories to a user query, and they are ranked by the system to be the first and the third; then the accuracy is computed in the following way: $s_{c1} = 1/(1+1-1) = 1$ and $s_{c2} = 1/(1+3-2) = 0.5$, such that the accuracy is $QC = (1 + 0.5)/2 = 0.75$. QC can be used to measure the accuracy of query categorization; the higher QC, the better the accuracy.

The query categorization method proposed in Section 8.5 is implemented and evaluated on a set of the 100 randomly sampled queries from the query log. Table 8.6 shows QC for the 5 different query examples and the average QC for those 100 queries in the test set, respectively. The average value of QC for the 100 queries indicates that the categorization results are satisfactory for most queries.

8.6.4 Search Relevance Results

Since the number of the video clips indexed in our experiments is very large, it is impossible to estimate the recall of the search for each query. In the experiments the evaluation is conducted using the following procedure. First, a

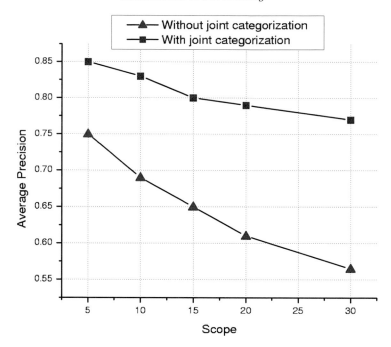

FIGURE 8.6: Comparisons of the search relevance with and without the joint categorization of query and video. The average precision vs. scope (top N results) curves. Reprint from [239] ©2006 ACM Press.

user submits a query and then the top three related categories obtained by the query categorization are returned along with the video results retrieved without using any categorization. Second, the user selects a category fitting his/her search query context. Third, the new search is refined to the video categorized to that category and the results are returned. The impact of the joint categorization of video and query to the search relevance is measured by computing the precisions for different numbers of results. Ten laymen are invited to test the system with and without the joint categorization function enabled. The relevance of the returned video clips is determined by the subjects. 223 queries are recorded, and the average precision-scope curves are plotted in Figure 8.6.

The figure indicates that the joint categorization of query and video by using this proposed approach improves the search relevance significantly (the precision is enhanced by 10%–20%).

It is noticeable that the video categorization is performed offline and the query categorization is performed online. The average time for categorizing a video clip is less than 0.1 second and the average time for categorizing a query is less than 0.05 second. The in-category search is very efficient; the average response time of the proposed prototype system for users' queries is less than 0.2 second. The high efficiency of the system makes it a practical search engine for the real-world video search on the Web.

8.7 Summary

This chapter demonstrates an example of concept discovery and mining in a large-scale video database by showcasing a developed video search engine prototype on the Web. We investigate several promising approaches to boosting the search relevance of a large-scale video search engine on the Web. We have developed a specialized video categorization framework which combines multiple classifiers based on different modalities. By learning users' querying histories and clicking logs, we have proposed an automatic query profile generation technique and have applied the profile to query categorization. A highly scalable prototype system is developed, which integrates the online query categorization and the offline video categorization. The naive Bayes with a mixture of multinomials, the maximum entropy, and the support vector machine categorization methods and the profile learning technique are evaluated on millions of the video clips on the Web. The evaluation of the developed system and user studies has indicated that the joint categorization of queries and video clips boosts the video search relevance and user search experience. The high efficiency of the proposed approaches is also demonstrated by the good responsiveness of the prototype system for the video search engine on the Web.

Chapter 9

Concept Discovery and Mining in an Audio Database

9.1 Introduction

In the last four chapters we have showcased the examples on knowledge discovery and data mining for imagery databases, video databases, as well as the multimedia databases combined with multiple modalities such as imagery combined with text and/or video combined with text that we have experienced in the commonly encountered multimedia data collections such as the Web. In this chapter, we focus on the topic of knowledge discovery and data mining in an audio database.

Audio data are substantially different from the data in other media types such as text, imagery, and video in the sense that they are essentially one-dimensional data. Like knowledge discovery in other types of multimedia databases as we have shown in the last few chapters, knowledge discovery in an audio database is also context-dependent and task-specific. Here in this chapter, we focus on concept discovery and mining in an audio database. Specifically, we address the problem of concept discovery and data mining in an audio database by showcasing an exemplar solution to audio classification and categorization of a typical audio database consisting of mixed types of audio data, including music, speech sounds, non-speech utterances, animal sounds, environmental sounds, as well as noise. Note that it is not uncommon that a typical audio database may contain up to millions of audio clips, resulting in audio classification and categorization as the very first and essential concept discovery and mining task before any more sophisticated knowledge discovery tasks may be demanded. On the other hand, given the scales of millions of audio clips in a typical audio database, it would become very tedious, if not impossible, to do a manual classification and categorization. Consequently, it is important that we study the audio classification and categorization problem as an example of concept discovery and mining in an audio database.

As an exemplar solution to audio data classification and categorization to be introduced and studied here in this chapter, we refer to the specific method by Lin et al [138]. The reasons why we pick up this method as an exemplar solution to study in this chapter are based on the following considerations. First, this work was published in 2005 and thus represents the state-of-the-art literature on audio data mining. Second, unlike many existing efforts in the audio data mining literature which mainly focus on a specific type of audio data mining such as speech data mining, this work is for classification and categorization of all different types of audio data. Third, classification and categorization represent one of the most important applications of audio data mining in current literature and also represent one of the most urgent needs in today's real-world applications of audio data mining. For reference purposes, we call this method LCTC method (for the last-name initials of all the four authors of this work).

This chapter is organized as follows. Section 9.2 gives a brief background description and literature review on the general audio mining as well as the specific related work on audio data classification and categorization. Section 9.3 introduces the specific features used in the LCTC method as well as how the features are extracted. Section 9.4 presents the specific solution to audio classification and categorization based on the LCTC method. Section 9.5 briefly discusses the experimental evaluations of the LCTC method. Finally, this chapter is concluded in Section 9.6.

9.2 Background and Related Work

Audio data represent sound information. As is discussed in Section 2.2.2, audio data are one-dimensional data; they may be plotted as a one-dimensional waveform. Depending upon the different sound sources, audio data may look differently. Figure 9.1 gives samples of different types of audio data. Clearly, from the waveforms of these samples, different types of audio data have different "shapes" in the plotting waveforms. For example, speech sound looks very different from music sound in Figure 9.1. Due to this difference, it is important that we select and use different audio features for different mining problems. For example, if the audio data mining problem is to classify and categorize the audio data, we need to ensure that the selected features are sufficiently discriminant for different types of audio data; on the other hand, if the audio data mining problem is related to speech recognition and retrieval, we need to ensure that the selected features respond well to speech data only.

Audio data mining may trace back to the early 1990s when audio data indexing and retrieval began to receive attention as the multimedia area started to develop. A notable example is the New Zealand digital libraries project [13, 219], where music melody data are stored in a digital library for indexing and retrieval. Since different types of audio data require different feature representations and thus may use different data mining techniques, and since in

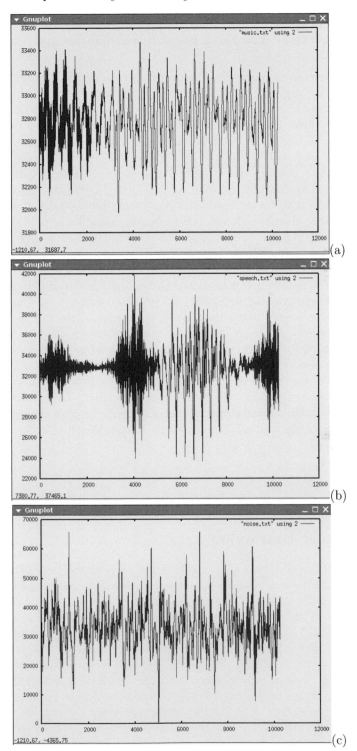

FIGURE 9.1: Samples of typical audio data. (a) Music sample. (b) Speech sample. (c) Noise sample.

this chapter we showcase an example on audio data classification and categorization, we give a brief review on the audio classification and categorization work.

Early work on audio data classification dated back to Wold et al [220], who developed the "Muscle Fish" database that was used later in the literature as a benchmark database [5]. Different features are used in audio classification literature. Examples include mel-frequency cepstral features [137], wavelet features [138, 44, 106], brightness and bandwidth features [137], subband power features [137, 106], and pitch features [44]. In terms of classification techniques, typically statistical techniques are used for classification and categorization. Noticeable examples include neural networks, hidden Markov models, and support vector machines. Zhang and Kuo [247] proposed a rule-based method to classify different types of audio data. Lu et al [141] proposed a two-step method for audio data classification; the first step is for speech and non-speech binary classification, and the K-nearest neighbor and the linear spectral-pairs vector quantization methods are used in this step; the second step uses a rule-based classification method to further classify the non-speech data into different types of audio data. Li proposed using the nearest-neighbor line method to classify audio data [137]. Guo and Li further improved the method by Li [137] by using support vector machines instead of the nearest neighbor lines [94]. Lin et al [138] proposed a new method for audio data classification and categorization based on Guo and Li's method by incorporating the wavelet features and by using the bottom-up SVMs, which is the LCTC method we introduce in this chapter. Recently, Ravindran et al [172] proposed using the Gaussian mixture model based classifier for audio classification, and Sainath et al used the extended Baum-Welch transformations for audio data classification [183].

9.3 Feature Extraction

The features used in the LCTC method include both the perceptual features and the transformation features specifically defined as the frequency cepstral coefficients. Before feature extraction, the original audio signal needs to be preprocessed to identify reliably all the nonsilent frames. The original audio signal is sampled at 8,000 Hz with 16-bit resolution. Each audio clip data stream is divided into frames. The frame length is set as 256 samples that correspond to 32 ms with a 192-sample (i.e., 75%) overlap between neighboring frames. Due to typical speech sounds' relatively low amplitude in the frequency spectrum contributed from the radiation effect of the sound from human lips, the energy enhancement technique is used at the high-frequency end in the spectrum [50]. This is achieved by using an enhancement filter defined as follows:

$$s'_n = s_n - 0.96s_{n-1}, \quad n = 1, \ldots, 255 \qquad (9.1)$$
$$s'_0 = s_0$$

where s_n and s'_n are the nth audio sample in a frame before and after, respectively, the application of the filter. The filtered audio signal is then further Hamming-weighted as follows:

$$s^h{}_n = s'_n h_n, \quad n = 0, \ldots, 255 \qquad (9.2)$$
$$h_n = 0.54 - 0.46 \cos(\tfrac{2\pi n}{255})$$

After this preprocessing, a frame is declared as a nonsilent frame if the frame satisfies the following constraint:

$$\sum_{n=0}^{255} (s^h{}_n)^2 > 400^2 \qquad (9.3)$$

where 400 is an experience threshold reported in [137, 94].

After obtaining all the nonsilent frames of the original audio signal after the preprocessing, we are ready to define and extract all the features designed for this specific classification and categorization problem. Essentially, the LCTC method uses both perceptual features and the transformation features. Specifically, the perceptual features are obtained by applying both the Fourier transform and the wavelet transform. The transformation features are obtained using the frequency cepstral coefficients (FCC). We describe the specific definitions for these features as follows.

- Subband power P_j: Three sections of the subband power are computed in the wavelet domain [144, 31]. Let ω be the half sampling frequency. The subband intervals are then $[0, \omega/8]$, $[\omega/8, \omega/4]$, and $[\omega/4, \omega/2]$, corresponding to the approximation and detail coefficients of the wavelet transform, respectively. The subband power is thus computed as

$$P_j = \sum_k z_j{}^2(k) \qquad (9.4)$$

where $z_j(k)$ is the corresponding approximation or detail coefficients of subband j.

- Pitch frequency f_p: The noise-robust wavelet based pitch detection method proposed by Chen and Wang [44] is used for pitch frequency extraction. The first stage of this method is to apply the wavelet transform with an aliasing compensation [144] to decompose the input signal into three subbands; this is followed by applying a modified spatial correlation function computed from the approximation signal obtained in the previous stage for pitch frequency extraction.

Table 9.1: List of the Extracted Features. Redrawn from [138].

Features		Type of transforms	Number of features
Perceptual features	Subband power P_j	Wavelet	3
	Pitch frequency f_p	Wavelet	1
	Brightness ω_c	Fourier	1
	Bandwidth B	Fourier	1
Frequency cepstral coefficient (FCC) c_n		Fourier	L

- Brightness ω_c: The brightness is defined as the frequency centroid of the Fourier trnasform and is computed as

$$\omega_c = \frac{\int_0^\infty u\|F(u)\|^2 du}{\int_0^\omega \|f(u)\|^2 du} \qquad (9.5)$$

- Bandwidth B: The bandwidth is defined as the square of the power-weighted average of the squared difference between the spectral components and the frequency centroid:

$$B = \sqrt{\frac{\int_0^\omega (u - \omega_c)^2 \|F(u)\|^2 du}{\int_0^\infty \|F(u)\|^2 du}} \qquad (9.6)$$

- Frequency Cepstral Coefficient (FCC) c_n: The FCC is defined as the L-order coefficients computed as

$$c_n = \sqrt{\frac{1}{128} \sum_{u=0}^{255} (\log_{10} F(u)) \cos \frac{n(u-0.5)\pi}{256}}, \qquad n = 1, \ldots, L \qquad (9.7)$$

Table 9.1 summarizes all the features defined and used in the LCTC method. Clearly, there are a total of $6 + L$ features. For each of the $6 + L$ features, the statistical mean and the standard deviation are computed, resulting in a total of $12 + 2L$ features. In addition, the pitch ratio, which is defined as the ratio of the number of the pitched frames to the total number of the frames in the signal, and the silence ratio, which is defined as the ratio of the number of the silent frames to the total number of the frames in the signal, are computed as the two further features. Consequently, a $14 + 2L$-dimensional feature vector is obtained.

In order to facilitate the subsequent classification, all the samples obtained as the $14 + 2L$-dimensional vectors need to be normalized. The normalization is performed in two steps. Given each sample represented as a $14 + 2L$-dimensional vector T_j, the first normalization step is to shift the vector in the

$14 + 2L$-dimensional space relative to the distribution center, i.e.,

$$T'_j = \frac{T_j - \mu_j}{\sigma_j} \tag{9.8}$$

where

$$\mu_j = \sum \frac{T_j}{N} \quad \sigma_j = \sum \frac{(T_j - \mu_j)^2}{N} \tag{9.9}$$

where N is the total number of the samples in a set (e.g., the training set). The second step is to further normalize the values of the $2L$ components in each of the sample vectors related to FCC, i.e.,

$$T''_j = \frac{T'_j}{m_j} \tag{9.10}$$

where m_j is the maximum of the absolute value for all the components of sample T'_j.

9.4 Classification Method

Since the goal is to classify and categorize the audio clips in an audio database, there are many classification methods in the literature that can be used. Specifically, the LCTC method elects to use support vector machines (SVMs) for the classification. In order to accommodate the potential classification errors, the soft-margin SVMs are used. In addition, both the RBF and Gaussian kernel functions are used in order to make a comparison regarding which is more effective in the classification using the same features. Since typically there are more than two classes available, the bottom-up binary tree approach is used for multiple class SVMs in the LCTC method. See Section 3.7 for a detailed discussion on the soft-margin SVMs, the different commonly used kernel functions, as well as the different approaches to extending the binary classes SVMs to the multiple-classes SVMs. Figure 9.2 shows an example of the reconstructed bottom-up binary tree after node 12 is removed.

9.5 Experimental Results

The LCTC method is evaluated using the publicly available Muscle Fish database [5]. The Muscle Fish database consists of 410 sound clips categorized into 16 classes. Table 9.2 lists the ground truth of the clip files in the Muscle

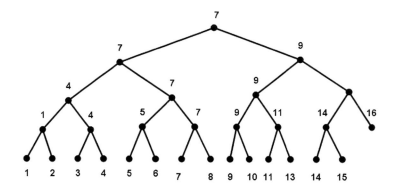

FIGURE 9.2: The reconstructed bottom-up binary tree after node 12 is removed.

Fish database. All the 410 clip files are sorted alphabetically in terms of the file names; the odd-numbered files in the sorted order are assigned to the training set; the remaining files are for the testing set. This results in 211 training files and 199 testing files.

As described in Section 9.3, each sample vector is represented as a normalized $14 + 2L$-dimensional vector, where the total number of samples, N, in Equation 9.9 is taken as the total number of the training samples, and the normalization parameters, μ_j, σ_j, and m_j in Equations 9.8 to 9.10 are all computed using the training data; these parameters are used in normalizing the samples in the testing data set, also.

As mentioned in Section 9.4, both RBF and Gaussian kernel functions are used for the comparison purpose. They are compared over a range of preselected values for the upper bound C and the variance σ^2, where the FCC level L varies from 1 to 99. This setting translates to a total of 144 pairs of C and σ^2 for each of the two kernel functions. For each combination of C and σ^2 values, E_m is defined as the least value of the errors, and L_m is defined as the specific FCC level L when the first E_m happens. Tables 9.3 and 9.4 document the comparison evaluations between the RBF and the Gaussian kernel functions. From these two tables, it is clear that the RBF kernel achieves a better accuracy than the Gaussian kernel in most settings. It is further observed that the RBF kernel function is more stable than the Gaussian kernel function when $C > 20$, suggesting that a larger value of

Table 9.2: Ground Truth of the Muscle Fish Database.

Class name	Number of clip files
Alto-trombone	13
Animals	9
Bells	7
Cello-bowed	47
Female	35
Laughter	7
Machines	11
Male	17
Oboe	32
Percussion	99
Telephone	17
Tubular-bells	20
Violin-bowed	45
Violin-pizz	40
Water	7

C has no additional benefit. Table 9.5 reports the performance comparison evaluations between the LCTC method and two state-of-the-art methods from the literature, the method developed by Guo and Li [94] (called the GL method for reference purposes) and the method developed by Li [137] (called the L method for reference purposes). Note that both the GL and the L methods use the same features, whereas the GL method uses SVMs for the classification and the L method uses the nearest feature line method (NFL), the nearest neighbor method (NN), the top 5 nearest neighbors method (5-NN), and the nearest center method (NC) for the classification. From the table, it is clear that the LCTC method outperforms both the GL and the L methods, and the GL method outperforms the L method.

Since it is shown that the RBF kernel outperforms the Gaussian kernel, further evaluations are reported using the RBF kernel function for the categorization with the bottom-up binary tree method for multiple-classes classification. Table 9.6 documents the evaluations. From these evaluations, it is clear that the LCTC method achieves 100% accuracy in the top 2 returns for most settings of C and σ^2 values. Further, high accuracy is achieved at many reasonably sized FCC levels, which demonstrates the promise and the effectiveness of the LCTC method. For all the six misclassified testing clip files, this is due to the fact that the sounds are very similar to the training sounds in other classes, even to the human ear.

Multimedia Data Mining

Table 9.3: Experimental results for the preselected values of C and σ^2 with the RBF kernel. Redrawn from [138].

E_m/L_m		C										
		1	5	10	20	30	40	50	60	70	80	90	100
	1	43 /3	38 /2	38 /2	38 /2	38 /2	38 /2	38 /2	38 /2	38 /2	38 /2	38 /2	38 /2
	5	40 /39	18 /11	18 /11	18 /11	18 /11	18 /11	18 /11	18 /11	18 /11	18 /11	18 /11	18 /11
	10	41 /54	12 /51	12 /51	12 /51	12 /51	12 /51	12 /51	12 /51	12 /51	12 /51	12 /51	12 /51
	20	60 /89	7 /54	7 /54	7 /54	7 /54	7 /54	7 /54	7 /54	7 /54	7 /54	7 /54	7 /54
	30	80 /99	9 /84	7 /54	7 /54	7 /54	7 /54	7 /54	7 /54	7 /54	7 /54	7 /54	7 /54
	40	91 /95	12 /86	7 /54	7 /54	7 /54	7 /54	7 /54	7 /54	7 /54	7 /54	7 /54	7 /54
σ^2	50	97 /95	14 /85	7 /54	7 /54	7 /54	7 /54	7 /54	7 /54	7 /54	7 /54	7 /54	7 /54
	60	102 /95	16 /82	7 /66	6 /80	6 /80	6 /80	6 /80	6 /80	6 /80	6 /80	6 /80	6 /80
	70	110 /82	17 /98	8 /83	6 /80	6 /80	6 /80	6 /80	6 /80	6 /80	6 /80	6 /80	6 /80
	80	114 /96	19 /84	11 /81	6 /80	6 /80	6 /80	6 /80	6 /80	6 /80	6 /80	6 /80	6 /80
	90	123 /87	19 /88	12 /87	38 /2	38 /2	38 /2	38 /2	38 /2	38 /2	38 /2	38 /2	38 /2
	100	127 /98	22 /90	13 /99	38 /2	38 /2	38 /2	38 /2	38 /2	38 /2	38 /2	38 /2	38 /2

Table 9.4: Experimental results for the preselected values of C and σ^2 with the Gaussian kernel. Redrawn from [138].

E_m/L_m		C										
		1	5	10	20	30	40	50	60	70	80	90	100
	1	35/9	20/11	20/11	19/11	20/8	19/11	20/11	20/11	19/11	20/11	20/11	20/11
	5	51/99	19/53	14/53	13/55	13/55	13/55	13/55	13/55	13/55	13/55	13/55	13/55
	10	73/89	20/96	15/96	14/96	14/96	14/96	14/96	14/96	14/96	14/96	14/96	14/96
	20	93/93	30/96	18/96	17/87	17/91	17/58	17/58	17/58	17/58	17/58	17/58	17/58
	30	101/77	35/95	22/96	16/95	16/95	16/95	16/95	16/95	16/95	16/95	16/95	16/95
	40	110/54	48/97	26/96	18/95	16/95	17/91	17/95	16/95	16/95	16/95	16/95	16/95
σ^2	50	120/81	58/98	31/96	19/97	18/95	16/95	17/91	17/95	16/95	16/95	16/95	16/95
	60	131/99	64/95	35/95	22/97	18/95	16/95	16/95	17/95	17/95	16/95	16/95	16/95
	70	135/95	72/99	39/96	23/96	19/95	18/95	16/95	17/95	17/95	17/95	16/95	16/95
	80	137/97	78/99	45/98	25/99	21/95	18/95	16/95	17/95	17/95	17/95	17/95	17/95
	90	142/99	86/99	52/98	28/96	22/97	19/96	17/96	16/96	17/96	18/89	18/89	18/89
	100	147/97	89/99	57/99	30/97	24/95	19/96	18/96	18/96	17/95	18/89	18/89	18/89

Table 9.5: Error rates (number of errors) comparison among the LCTC, GL, and L methods (where NPC-L means the number of errors/$199 \times 100\%$, and PercCepsL means the number of errors/$198 \times 100\%$). Redrawn from [138].

Method	LCTC		GL	L			
Feature Set	NPC-L		PercCepsL	PercCepsL			
Classifier and Kernel	SVM RBF $C=30$, $\sigma^2=60$	SVM Gaussian $C=100$, $\sigma^2=5$	SVM RBF $C=200$, $\sigma^2=6$	NFL	NN	5-NN	NC
$L=5$	11.6% (23)	12.6% (25)	12.6% (25)	12.1% (24)	17.7% (35)	21.2% (42)	43.4% (86)
$L=8$	9.5% (19)	10.6% (21)	8.1% (16)	9.6% (19)	13.1% (26)	22.2% (44)	38.9% (77)
$L=60$	3.5% (6)	9.5% (19)	10.1% (20)	12.1% (24)	15.7% (31)	20.7% (41)	32.8% (65)

Table 9.6: Categorization errors in the top 2 returns using the RBF kernel function for the pre-selected values of C and σ^2. Redrawn from [138].

	E_m/L_m	\multicolumn{12}{c}{C}

	E_m/L_m	1	5	10	20	30	40	50	60	70	80	90	100
	1	23/1	19/4	19/4	19/4	19/4	19/4	19/4	19/4	19/4	19/4	19/4	19/4
	5	11/29	3/29	3/29	3/29	3/29	3/29	3/29	3/29	3/29	3/29	3/29	3/29
	10	17/37	0/46	0/46	0/46	0/46	0/46	0/46	0/46	0/46	0/46	0/46	0/46
	20	22/98	0/38	0/37	0/37	0/37	0/37	0/37	0/37	0/37	0/37	0/37	0/37
	30	28/96	0/49	0/29	0/29	0/29	0/29	0/29	0/29	0/29	0/29	0/29	0/29
	40	37/96	1/77	0/29	0/29	0/29	0/29	0/29	0/29	0/29	0/29	0/29	0/29
σ^2	50	49/99	3/81	0/40	0/29	0/29	0/29	0/29	0/29	0/29	0/29	0/29	0/29
	60	66/91	3/97	0/46	0/29	0/29	0/29	0/29	0/29	0/29	0/29	0/29	0/29
	70	76/99	4/99	0/50	0/29	0/29	0/29	0/29	0/29	0/29	0/29	0/29	0/29
	80	79/85	5/76	0/68	0/29	0/29	0/29	0/29	0/29	0/29	0/29	0/29	0/29
	90	84/93	5/95	1/80	0/33	0/29	0/29	0/29	0/29	0/29	0/29	0/29	0/29
	100	90/76	8/90	2/87	0/39	0/29	0/29	0/29	0/29	0/29	0/29	0/29	0/29

9.6   Summary

In this chapter, we focus on audio data classification and categorization as a case study for the application of audio data mining. We first give a brief background introduction and review on the audio data classification literature. Then we give a specific example of the audio data classification literature by introducing the method recently developed by Lin et al [138]. We have introduced the specific features used in the method as well as the specific method itself. We have also reported the experimental evaluations of this method in comparison with the peer methods in the literature. We hope that this case study gives readers the flavor of the state-of-the-art audio data classification and categorization literature.

References

[1] http://wordnet.princeton.edu/

[2] http://www.corel.com/

[3] http://svmlight.joachims.org/svm_perf.html

[4] http://www.mpeg.org/

[5] http://www.musclefish.com/cbrdemo.html

[6] G. Adomavicius and A. Tuzhilin. Toward the next generation of recommender systems: a survey of the state-of-the-art and possible extensions. *IEEE Transactions on Knowledge and Data Engineering*, 17(6):734–749, 2005.

[7] M. Aizerman, E. Braverman, and L. Rozonoer. Theoretical foundations of the potential function method in pattern recognition learning. *Automation and Remote Control*, (A25):821–837, 1964.

[8] R. A. Aliev and R. R. Aliev. *Soft Computing and Its Applications*. World Scientific, September 2001.

[9] Y. Altun, I. Tsochantaridis, and T. Hofmann. Hidden Markov support vector machines. In *Proc. ICML*, Washington DC, August 2003.

[10] A. Amir, J. Argillander, M. Campbell, A. Haubold, G. Iyengar, S. Ebadollahi, and F. Kang. IBM research TRECVID-2005 video retrieval system. In *NIST TRECVID-2005 Workshop*, Gaithersburg, MD, November 2005.

[11] P. Auer. On learning from multi-instance examples: empirical evaluation of a theoretical approach. In *Proc. ICML*, 1997.

[12] R. Baeza-Yates and B. Ribeiro-Neto. *Modern Information Retrieval*. Addison-Wesley, June 1999.

[13] D. Bainbridge, M. Dewsnip, and I. Witten. Searching digital music libraries. *Information Processing and Management*, 41(1):41–56, 2005.

[14] K. Barnard, P. Duygulu, N. d.Freitas, D. Blei, and M. I. Jordan. Matching words and pictures. *Journal of Machine Learning Research*, 3:1107–1135, 2003.

[15] K. Barnard and D. Forsyth. Learning the semantics of words and pictures. In *The International Conference on Computer Vision*, volume II, pages 408–415, 2001.

[16] A. G. Barto and P. Anandan. Pattern recognizing stochastic learning automata. *IEEE Trans. Systems, Man, and Cybernetics*, (15):360–375, 1985.

[17] M.J. Beal and Z. Ghahramani. The infinite hidden Markov model. In *Proc. NIPS*, 2002.

[18] M. Belkin, P. Niyogi, and V. Sindhwani. Manifold regularization: A geometric framework for learning from labeled and unlabeled examples. *Journal of Machine Learning Research*, 7:2399–2434, 2006.

[19] D. Blackwell and J.B. MacQueen. Ferguson distribution via polya urn schemes. *The Annal of Statistics*, 1:353–355, 1973.

[20] V. Blanz, B. Schölkopf, H. Bülthoff, C. Burges, V. Vapnik, and T. Vetter. Comparison of view-based object recognition algorithms using realistic 3d models. In *Artificial Neural Networks 96 Proceedings*, pages 251–256, Berlin, Germany, 1996.

[21] D. Blei and M. Jordan. Modeling annotated data. In *The 26th International Conference on Research and Development in Information Retrieval (SIGIR)*, 2003.

[22] D. Blei, A. Ng, and M. Jordan. Dirichlet allocation models. In *The International Conference on Neural Information Processing Systems*, 2001.

[23] D. M. Blei, A. Y. Ng, and M. I. Jordan. Latent dirichlet allocation. *Journal of Machine Learning Research*, (3):993–1022, 2003.

[24] D.M. Blei and M.I. Jordan. Variational methods for the Dirichlet process. In *Proc. ICML*, 2004.

[25] G. Blom. *Probability and Statistics: Theory and Applications*. Springer Verlag, London, U.K., 1989.

[26] A. Blum and T. Mitchell. Combining labeled and unlabeled data with co-training. In *Proc. Workshop on Computational Learning Theory*. Morgan Kaufman Publishers, 1998.

[27] B. E. Boser, I. M. Guyon, and V. N. Vapnik. A training algorithm for optimal margin classifiers. In *The 5th Annual ACM Workshop on COLT*, pages 144–152, Pittsburgh, PA, 1992.

[28] S. Boyd and L. Vandenberghe. *Convex Optimization*. Cambridge University Press, 2004.

[29] R. Brachman and H. Levesque. *Readings in Knowledge Representation.* Morgan Kaufman, 1985.

[30] A. Brink, S. Marcus, and V. Subrahmanian. Heterogeneous multimedia reasoning. *Computer*, 28(9):33-39, 1995.

[31] C.S. Burrus, R.A. Gopinath, and H. Guo. *Introduction to Wavelets and Wavelet Transforms.* Prentice-Hall, 1998.

[32] D. Cai, S. Yu, J.-R. Wen, and W.-Y. Ma. VIPS: A vision-based page segmentation algorithm. Microsoft Technical Report (MSR-TR-2003-79), 2003.

[33] L. Cao and L. Fei-Fei. Spatially coherent latent topic model for concurrent object segmentation and classification. In *Proc. ICCV*, 2007.

[34] P. Carbonetto, N. d. Freitas, and K. Barnard. A statistical model for general contextual object recogniton. In *The 8th European Conference on Computer Vision*, 2004.

[35] P. Carbonetto, N. d. Freitas, P. Gustafson, and N. Thompson. Bayesian feature weighting for unsupervised learning, with application to object recognition. In *The 9th International Workshop on Artificial Intelligence and Statistics*, 2003.

[36] C. Carson, S. Belongie, H. Greenspan, and J. Malik. Blobworld: Image segmentation using expectation-maximization and its application to image querying. *IEEE Trans. on PAMI*, 24(8):1026–1038, 2002.

[37] C. Carson, M. Thomas, S. Belongie, J.M. Hellerstein, and J. Malik. Blobworld: A system for region-based image indexing and retrieval. In *The 3rd Int'l Conf. on Visual Information System Proceedings*, pages 509–516, Amsterdam, Netherlands, June 1999.

[38] K.R. Castleman. *Digital Image Processing.* Prentice Hall, Upper Saddle River, NJ, 1996.

[39] U. Cetintemel, M. J. Franklin, and C. L. Giles. Self-adaptive user profiles for large-scale data delivery. In *ICDE*, 2000.

[40] E. Chang, K. Goh, G. Sychay, and G. Wu. CBSA: Content-based soft annotation for multimodal image retrieval using Bayes point machines. *IEEE Trans. on Circuits and Systems for Video Technology*, 13(1):26–38, January 2003.

[41] S. F. Chang, J. R. Smith, M. Beigi, and A. Benitez. Visual information retrieval from large distributed online repositories. *Comm. ACM*, 40(2):63–67, 1997.

[42] W. Chang, G. Sheikholeslami, J. Wang, and A. Zhang. Data resource selection in distributed visual information systems. *IEEE Trans. on Knowledge and Data Engineering*, 10(6):926–946, Nov./Dec. 1998.

[43] O. Chapelle, P. Haffner, and V.N. Vapnik. Support vector machines for histogram-based image classification. *IEEE Trans. on Neural Networks*, 10(5):1055–1064, September 1999.

[44] S.-H. Chen and J.-F. Wang. Noise-robust pitch detection method using wavelet transform with aliasing compensation. *Proceedings of IEE Vision, Image Signal Processing*, 149(6):327–334, 2002.

[45] S.F. Chen and R. Rosenfeld. A Gaussian prior for smoothing maximum entropy models. Technical report, CMU, 1999.

[46] Y. Chen, J. Bi, and J.Z. Wang. MILES: multiple-instance learning via embedded instance selection. *IEEE Trans. PAMI*, 28(12), 2006.

[47] Y. Chen and J.Z. Wang. A region-based fuzzy feature matching approach to content-based image retrieval. *IEEE Trans. on PAMI*, 24(9):1252–1267, 2002.

[48] Y. Chen, J.Z. Wang, and R. Krovetz. Content-based image retrieval by clustering. In *The 5th ACM SIGMM International Workshop on Multimedia Information Retrieval*, pages 193–200, Berkeley, CA, November 2003.

[49] D.G. Childers, D.P. Skinner, and R.C. Kemerait. The cepstrum: A guide to processing. *Proceedings of the IEEE*, 65(10):1428–1443, 1977.

[50] W.C. Chu. *Speech Coding Algorithms*. Wiley, 2003.

[51] G. Cooper. Computational complexity of probabilistic inference using Bayesian belief networks (research note). *Artificial Intelligence*, (42):393–405, 1990.

[52] G. Cooper and E. Herskovits. A Bayesian method for the induction of probabilistic networks from data. *Machine Learning*, (9):309–347, 1992.

[53] C. Cortes and V. Vapnik. Support-vector networks. *Machine Learning*, 20(3):273–297, 1995.

[54] P. Dagum and M. Luby. Approximating probabilistic reasoning in Bayesian belief networks is NP-Hard. *Artificial Intelligence*, 60(1):141–153, 1993.

[55] H. Daume III and D. Marcu. Learning as search optimization: Approximate large margin methods for structured prediction. In *Proc. ICML*, Bonn, Germany, August 2005.

[56] S. Deerwester, S. Dumais, G. Furnas, T. Landauer, and R. Harshman. Indexing by latent semantic analysis. *Journal of the American Society for Information Science*, 41:391–407, 1990.

[57] A. Demiriz, K.P. Bennett, and J. Shawe-Taylor. Linear programming boosting via column generation. *Kluwer Machine Learning*, (46):225–254, 2002.

[58] A. Dempster, N. Laird, and D. Rubin. Maximum likelihood from incomplete data via the EM algorithm. *Journal of the Royal Statistical Society, Series B*, 39(1):1–38, 1977.

[59] T.G. Dietterich, R.H. Lathrop, and T. Lozano-Perez. Solving the multiple instance problem with axis-parallel rectangles. *Artificial Intelligence*, 89:31–71, 1997.

[60] W. R. Dillon and M. Goldstein. *Multivariate Analysis, Methods and Applications*. John Wiley and Sons, New York, 1984.

[61] N. Dimitrova, L. Agnihotri, and G. Wei. Video classification based on HMM using text and faces. In *European Conference on Signal Processing*, Finland, September 2000.

[62] C. Djeraba, editor. *Multimedia Mining — A Highway to Intelligent Multimedia Document*. Kluwer Academic Publishers, 2002.

[63] C. Djeraba. Association and content-based retrieval. *IEEE Transaction on Knowledge and Data Engineering*, 15(1):118–135, January 2003.

[64] C. Domingo and O. Watanabe. Madaboost: A modification of adaboost. In *Proc. 13th Annu. Conference on Comput. Learning Theory*, pages 180–189, Morgan Kaufmann, San Francisco, 2000.

[65] H. Drucker, C.J.C. Burges, L. Kaufman, A. Smola, and V. Vapnik. Support vector regression machines. In *Advances in Neural Information Processing Systems 9, NIPS 1996*, pages 156–161, 1997.

[66] R. O. Duda and P. E. Hart. *Pattern Classification and Scene Analysis*. John Wiley and Sons, New York, 1973.

[67] R.O. Duda, P.E. Hart, and D.G. Stork. *Pattern Classification (2nd ed.)*. John Wiley and Sons, 2001.

[68] F. Dufaux. Key frame selection to represent a video. In *IEEE International Conference on Image Processing*, 2000.

[69] M.H. Dunham. *Data Mining, Introductory and Advanced Topics*. Prentice Hall, Upper Saddle River, NJ, 2002.

[70] P. Duygulu, K. Barnard, J. F. G. d. Freitas, and D. A. Forsyth. Object recognition as machine translation: Learning a lexicon for a fixed image vocabulary. In *The 7th European Conference on Computer Vision*, volume IV, pages 97–112, Copenhagen, Denmark, 2002.

[71] C. Faloutsos. *Searching Multimedia Databases by Content*. Kluwer Academic Publishers, 1996.

[72] C. Faloutsos, R. Barber, M. Flickner, J. Hafner, W. Niblack, D. Petkovic, and W. Equitz. Efficient and effective querying by image

content. *Journal of Intelligent Information Systems*, 3(3/4):231–262, 1994.

[73] L. Fei-Fei and P. Perona. A Bayesian hierarchical model for learning natural scene categories. In *Proc. CVPR*, pages 524–531, 2005.

[74] L. Fei-Fei and P. Perona. One-shot learning of object categories. *IEEE Trans. PAMI*, 28(4):594–611, 2006.

[75] S. L. Feng, R. Manmatha, and V. Lavrenko. Multiple Bernoulli relevance models for image and video annotation. In *The International Conference on Computer Vision and Pattern Recognition*, Washington, DC, June, 2004.

[76] R. Fergus, L. Fei-Fei, P. Perona, and A. Zisserman. Learning object categories from Google's image search. In *Proc. ICCV*, 2005.

[77] T. Ferguson. A Bayesian analysis of some non-parametric problems. *The Annal of Statistics*, 1:209–230, 1973.

[78] S. Fischer, R. Lienhart, and W. Effelsberg. Automatic recognition of film genres. In *The 3rd ACM International Conference on Multimedia*, San Francisco, CA, 1995.

[79] G. Fishman. *Monte Carlo Concepts, Algorithms and Applications*. Springer Verlag, 1996.

[80] M. Flickner, H.S. Sawhney, J. Ashley, Q. Huang, B. Dom, M. Gorkani, J. Hafner, D. Lee, D. Petkovic, D. Steele, and P. Yanker. Query by image and video content: The QBIC system. *IEEE Computer*, 28(9):23–32, September 1995.

[81] Y. Freund. Boosting a weak learning algorithm by majority. In *Proceedings of the Third Annual Workshop on Computational Learning Theory*, 1990.

[82] Y. Freund. An adaptive version of the boost by majority algorithm. *Machine Learning*, 43(3):293–318, 2001.

[83] Y. Freund and R.E. Schapire. A decision-theoretic generalization of online learning and an application to boosting. *Journal of Computer and System Sciences*, (55), 1997.

[84] Y. Freund and R.E. Schapire. Large margin classification using the perceptron algorithm. In *Machine Learning*, volume 37, 1999.

[85] J.H. Friedman. Stochastic gradient boosting. *Comput. Stat. Data Anal.*, 38(4):367–378, 2002.

[86] K. Fukunaga. *Introduction to Statistical Pattern Recognition (Second Edition)*. Academic Press, 1990.

[87] B. Furht, editor. *Multimedia Systems and Techniques*. Kluwer Academic Publishers, 1996.

[88] A. Gersho. Asymptotically optimum block quantization. *IEEE Trans. on Information Theory*, 25(4):373–380, 1979.

[89] M. Girolami and A. Kaban. On an equivalence between pLSI and LDA. In *SIG IR 2003*, 2003.

[90] Y. Gong and W. Xu. *Machine Learning for Multimedia Content Analysis*. Springer, 2007.

[91] H. Greenspan, G. Dvir, and Y. Rubner. Context dependent segmentation and matching in image databases. *Journal of Computer Vision and Image Understanding*, 93:86–109, January 2004.

[92] H. Greenspan, J. Goldberger, and L. Ridel. A continuous probabilistic framework for image matching. *Journal of Computer Vision and Image Understanding*, 84(3):384–406, December 2001.

[93] A. Grossmann and J. Morlet. Decomposition of hardy functions into square integrable wavelets of constant shape. *SIAM Journal on Mathematical Analysis*, 15(4), 1984.

[94] G. Guo and S.Z. Li. Content-based audio classification and retrieval by support vector machines. *IEEE Transactions on Neural Networks*, 14(1):209–215, 2003.

[95] Z. Guo, Z. Zhang, E.P. Xing, and C. Faloutsos. Enhanced max margin learning on multimodal data mining in a multimedia database. In *Proc. ACM International Conference on Knowledge Discovery and Data Mining*, 2007.

[96] Z. Guo, Z. Zhang, E.P. Xing, and C. Faloutsos. Semi-supervised learning based on semiparametric regularization. In *Proc. SIAM International Conference on Data Mining*, 2008.

[97] J. Han and M. Kamber. *Data Mining — Concepts and Techniques*. Morgan Kaufmann, 2nd edition, 2006.

[98] A.G. Hauptmann and M.G. Christel. Successful approaches in the TREC video retrieval evaluations. In *the 12th Annual ACM International Conference on Multimedia*, pages 668–675, New York City, NY, 2004.

[99] A.G. Hauptmann, R. Jin, and T.D. Ng. Video retrieval using speech and image information. In *Electronic Imaging Conference (EI'03), Storage and Retrieval for Multimedia Databases*, 2003.

[100] P. Hayes. The logic of frames. In R. Brachman and H. Levesque, editors, *Readings in Knowledge Representation*. Morgan Kaufmann, pages 288–295, 1979.

[101] T. Hofmann. Unsupervised learning by probabilistic latent semantic analysis. *Machine Learning*, 42(1):177–196, 2001.

[102] T. Hofmann and J. Puzicha. Statistical models for co-occurrence data. *AI Memo*, 1625, 1998.

[103] T. Hofmann, J. Puzicha, and M. I. Jordan. Unsupervised learning from dyadic data. In *The International Conference on Neural Information Processing Systems*, 1996.

[104] F. Hoppner, F. Klawonn, R. Kruse, and T. Runkler. *Fuzzy Cluster Analysis: Methods for Classification, Data Analysis and Image Recognition*. John Wiley & Sons, New York, 1999.

[105] B.K.P. Horn. *Robot Vision*. MIT Press and McGraw-Hill, 1986.

[106] C.-T. Hsieh and Y.-C. Wang. Robust speech features based on wavelet transform with application to speaker identification. *Proceedings of IEE Vision, Image Signal Processing*, 149(2):108–114, 2002.

[107] C.C. Hsu, W.W. Chu, and R.K. Raira. A knowledge-based approach for retrieving images by content. *IEEE Transactions on Knowledge and Data Engineering*, 8(4):522–532, August 1996.

[108] http://www-nlpir.nist.gov/projects/trecvid/. Digital video retrieval at NIST: TREC video retrieval evaluation 2001–2004, 2004.

[109] M.K. Hu. Visual pattern recognition by moment invariants. In J.K. Aggarwal, R.O. Duda, and A. Rosenfeld, editors, *Computer Methods in Image Analysis*. IEEE Computer Society Press, 1977.

[110] J. Huang, R. Kumar, and R. Zabih. An automatic hierarchical image classification scheme. In *The Sixth ACM Int'l Conf. Multimedia Proceedings*, 1998.

[111] J. Huang, Z. Liu, and Y. Wang. Joint video scene segmentation and classification based on hidden Markov model. In *IEEE International Conference on Multimedia and Expo (ICME)*, New York, NY, July 2000.

[112] J. Huang, S.R. Kumar, M. Mitra, W.-J. Zhu, and R. Zabih. Image indexing using color correlograms. In *IEEE Int'l Conf. Computer Vision and Pattern Recognition Proceedings*, Puerto Rico, 1997.

[113] R. Jain. Infoscopes: Multimedia information systems. In B. Furht, editor, *Multimedia Systems and Techniques*. Kluwer Academic Publishers, 1996.

[114] R. Jain. Content-based multimedia information management. In *Int'l Conf. Data Engineering Proceedings*, pages 252–253, 1998.

[115] R. Jain, R. Kasturi, and B.G. Schunck. *Machine Vision*. MIT Press and McGraw-Hill, 1995.

[116] E.T. Jaynes. Information theory and statistical mechanics. *The Physical Review*, 108:171–190, 1957.

[117] J. Jeon, V. Lavrenko, and R. Manmatha. Automatic image annotation and retrieval using cross-media relevance models. In *The 26th Annual International ACM SIGIR Conference on Research and Development in Information Retrieval*, 2003.

[118] F. Jing, M. Li, H.-J. Zhang, and B. Zhang. An effective region-based image retrieval framework. In *ACM Multimedia Proceedings*, Juan-les-Pins, France, December 2002.

[119] F. Jing, M. Li, H.-J. Zhang, and B. Zhang. An efficient and effective region-based image retrieval framework. *IEEE Trans. on Image Processing*, 13(5), May 2004.

[120] T. Joachims. Training linear SVMs in linear time. In *KDD 2006*, Philadelphia, PA, 2006.

[121] R. L. Kasyap, C. C. Blaydon, and K. S. Fu. Stochastic approximation. In K. S. Fu and J. M. Mendel, editors, *Adaptation, Learning, and Pattern Recognition Systems: Theory and Applications*. Academic Press, 1970.

[122] J. Kautsky, N. K. Nichols, and D. L. B. Jupp. Smoothed histogram modification for image processing. *CVGIP: Image Understanding*, 26(3):271–291, June 1984.

[123] M. Kearns. Thoughts on hypothesis boosting. Unpublished manuscript, 1988.

[124] S.S. Keerthi and D. DeCoste. A modified finite Newton method for fast solution of large scale linear SVMs. *Journal of Machine Learning Research*, 2005(6):341–361, 2005.

[125] S. Kendal and M. Creen. *An Introduction to Knowledge Engineering*. Springer, 2007.

[126] K.L. Ketner and H. Putnam. *Reasoning and the Logic of Things*. Harvard University Press, 1992.

[127] J. Kittler, M. Hatef, R. P. W. Duin, and J. Mates. On combining classifiers. *IEEE Transactions on Pattern Analysis and Machine Intelligence*, 20(3), 1998.

[128] G.J. Klir, U.H. St Clair, and B. Yuan. *Fuzzy Set Theory: Foundations and Applications*. Prentice Hall, 1997.

[129] T. Kohonen. *Self-Organizing Maps*. Springer, Berlin, Germany, 2001.

[130] T. Kohonen, S. Kaski, K. Lagus, J. Salojärvi, J. Honkela, V. Paatero, and A. Saarela. Self organization of a massive document collection. *IEEE Trans. on Neural Networks*, 11(3):1025–1048, May 2000.

[131] M. Koster. Alweb: Archie-like indexing in the web. *Computer Networks and ISDN Systems*, 27(2):175–182, 1994.

[132] N. Krause and Y. Singer. Leveraging the margin more carefully. In *Proceedings of the International Conference on Machine Learning (ICML)*, 2004.

[133] N. Kwak and C.-H. Choi. Input feature selection by mutual information based on parzen window. *IEEE Transactions on Pattern Analysis and Machine Intelligence*, 24(12):1667–1671, 2002.

[134] V. Lavrenko, R. Manmatha, and J. Jeon. A model for learning the semantics of pictures. In *The International Conference on Neural Information Processing Systems (NIPS'03)*, 2003.

[135] D.D. Lee and H.S. Seung. Algorithms for non-negative matrix factorization. In *Proc. NIPS*, pages 556–562, 2000.

[136] J. Li and J.Z. Wang. Automatic linguistic indexing of pictures by a statistical modeling approach. *IEEE Trans. on PAMI*, 25(9), September 2003.

[137] S.Z. Li. Content-based audio classification and retrieval using the nearest feature line method. *IEEE Transactions on Speech and Audio Processing*, 8(5):619–625, 2000.

[138] C.-C. Lin, S.-H. Chen, T.-K. Truong, and Y. Chang. Audio classification and categorization based on wavelets and support vector machine. *IEEE Transactions on Speech and Audio Processing*, 13(5):644–651, 2005.

[139] W.-H. Lin and A. Hauptmann. News video classification using SVM-based multimodal classifiers and combination strategies. In *ACM Multimedia*, Juan-les-Pins, France, 2002.

[140] P. Lipson, E. Grimson, and P. Sinha. Configuration based scene classification and image indexing. In *The 16th IEEE Conf. on Computer Vision and Pattern Recognition Proceedings*, pages 1007–1013, 1997.

[141] L. Lu, H.-J. Zhang, and H. Jiang. Content analysis for audio classification and segmentation. *IEEE Transactions on Speech and Audio Processing*, 10(7):504–516, 2002.

[142] W. Y. Ma and B. Manjunath. Netra: A toolbox for navigating large image databases. In *IEEE Int'l Conf. on Image Processing Proceedings*, pages 568–571, Santa Barbara, CA, 1997.

[143] W.Y. Ma and B. S. Manjunath. A comparison of wavelet transform features for texture image annotation. In *International Conference on Image Processing*, pages 2256–2259, 1995.

[144] S. Mallat. *A Wavelet Tour of Signal Processing*. Academic Press, 1998.

[145] B. S. Manjunath and W. Y. Ma. Texture features for browsing and retrieval of image data. *IEEE Trans. on Pattern Analysis and Machine Intelligence*, 18(8), August 1996.

[146] O. Maron and T. Lozano-Perez. A framework for multiple instance learning. In *Proc. NIPS*, 1998.

[147] L. Mason, J. Baxter, P. Bartlett, and M. Frean. Boosting algorithms as gradient descent. In *Proceedings of Advances in Neural Information Processing Systems 12*, pages 512–518, MIT Press, 2000.

[148] E. Mayoraz and E. Alpaydin. Support vector machines for multi-class classification. In *IWANN (2)*, pages 833–842, 1999.

[149] A. McGovern and D. Jensen. Identifying predictive structures in relational data using multiple instance learning. In *Proc. ICML*, 2003.

[150] G. Mclachlan and K. E. Basford. *Mixture Models*. Marcel Dekker, Inc., Basel, NY, 1988.

[151] S.W. Menard. *Applied Logistic Regression Analysis*. Sage Publications Inc, 2001.

[152] M. Minsky. A framework for representing knowledge. In P.H. Winston, editor, *The Psychology of Computer Vision*. McGraw-Hill, 1975.

[153] T.M. Mitchell. *Machine Learning*. McGraw-Hill, 1997.

[154] B. Moghaddam, Q. Tian, and T.S. Huang. Spatial visualization for content-based image retrieval. In *The International Conference on Multimedia and Expo 2001*, 2001.

[155] F. Monay and D. Gatica-Perez. PLSA-based image auto-annotation: constraining the latent space. In *Proc. ACM Multimedia*, 2004.

[156] Y. Mori, H. Takahashi, and R. Oka. Image-to-word transformation based on dividing and vector quantizing images with words. In *The First International Workshop on Multimedia Intelligent Storage and Retrieval Management*, 1999.

[157] K. S. Narenda and M. A. Thathachar. Learning automata — a survey. *IEEE Trans. Systems, Man, and Cybernetics*, (4):323–334, 1974.

[158] R. Neal. Markov chain sampling methods for Dirichlet process mixture models. *Journal of Computational and Graphical Statistics*, 9:249–265, 2000.

[159] K. Nigam, J. Lafferty, and A. McCallum. Using maximum entropy for text classification. In *IJCAI-99 Workshop on Machine Learning for Information Filtering*, pages 61–67, 1999.

[160] A.V. Oppenheim, A.S. Willsky, and I.T. Young. *Signals and Systems*. Prentice-Hall, 1983.

[161] M. Opper and D. Haussler. Generalization performance of Bayes optimal prediction algorithm for learning a perception. *Physics Review Letters*, (66):2677–2681, 1991.

[162] E. Osuna, R. Freund, and F. Girosi. An improved training algorithm for support vector machines. In *Proc. of IEEE NNSP'97*, Amelia Island, FL, September 1997.

[163] S.K. Pal, A. Ghosh, and M.K. Kundu. *Soft Computing for Image Processing*. Physica-Verlag, 2000.

[164] G. Pass and R. Zabih. Histogram refinement for content-based image retrieval. In *IEEE Workshop on Applications of Computer Vision*, Sarasota, FL, December 1996.

[165] M. Pazzani and D. Billsus. Learning and revising user profiles: the identification of interesting web sites. *Machine Learning*, pages 313–331, 1997.

[166] A. Pentland, R. W. Picard, and S. Sclaroff. Photobook: Tools for content-based manipulation of image databases. In *SPIE-94 Proceedings*, pages 34–47, 1994.

[167] V.A. Petrushin and L. Khan, editors. *Multimedia Data Mining and Knowledge Discovery*. Springer, 2006.

[168] S.D. Pietra, V.D. Pietra, and J. Lafferty. Inducing features of random fields. *IEEE Transactions on Pattern Analysis and Machine Intelligence*, 19(4), 1997.

[169] J. Platt. Fast training of support vector machines using sequential minimal optimization. In B. Schlkopf, C. Burges, and A. Smola, editors, *Advances in Kernel Methods — Support Vector Learning*. MIT Press, 1998.

[170] M. Pradham and P. Dagum. Optimal Monte Carlo estimation of belief network inference. In *Proceedings of the Conference on Uncertainty in Artificial Intelligence*, pages 446–453, 1996.

[171] A. L. Ratan and W. E. L. Grimson. Training templates for scene classification using a few examples. In *IEEE Workshop on Content-Based Access of Image and Video Libraries Proceedings*, pages 90–97, 1997.

[172] S. Ravindran, K. Schlemmer, and D.V. Anderson. A physiologically inspired method for audio classification. *EURASIP Journal on Applied Signal Processing*, 2005(1):1374–1381, 2005.

[173] J.D.M. Rennie, L. Shih, J. Teevan, and D.R. Karger. Tackling the poor assumptions of naive Bayes text classifiers. In *The 20th International Conference on Machine Learning (ICML'03)*, Washington, DC, 2003.

[174] J. Rissanen. Modelling by shortest data description. *Automatica*, 14:465–471, 1978.

[175] J. Rissanen. *Stochastic Complexity in Statistical Inquiry*. World Scientific, 1989.

[176] J. J. Rocchio Jr. Relevance feedback in information retrieval. In *The SMART Retrieval System — Experiments in Automatic Document Processing*, pages 313–323. Prentice Hall, Inc., Englewood Cliffs, NJ, 1971.

[177] R. Rosenfeld. *Adaptive statistical language modeling: A maximum entropy approach*. Ph.D. dissertation, Carnegie Mellon Univ., Pittsburgh, PA, 1994.

[178] Y. Rui, T. S. Huang, S. Mehrotra, and M. Ortega. A relevance feedback architecture in content-based multimedia information retrieval systems. In *IEEE Workshop on Content-based Access of Image and Video Libraries, in conjunction with CVPR'97*, pages 82–89, June 1997.

[179] D. E. Rummelhart, G. E. Hilton, and R. J. Williams. Learning internal representations by errors propagation. In D. E. Rummelhart, J. L. McClelland, and the PDP Research Group, editors, *Parallel Distributed Processing: Explorations in the Microstructure of Cognition, Volume 1: Foundations*. MIT Press, 1986.

[180] D. E. Rummelhart, G. E. Hilton, and R. J. Williams. Learning internal representations by back propagating errors. *Nature*, (323):533–536, 1986.

[181] B. Russell, A. Efros, J. Sivic, W. Freeman, and A. Zisserman. Using multiple segmentations to discover objects and their extent in image collections. In *Proc. CVPR*, 2006.

[182] S. Russell and P. Norvig. *Artificial Intelligence: A Modern Approach*. Prentice Hall, Upper Saddle River, NJ, 1995.

[183] T.N. Sainath, V. Zue, and D. Kanevsky. Audio classification using extended Baum-Welch transformations. In *Proc. of International Conference on Audio and Speech Signal Processing*, 2007.

[184] G. Salton. Developments in automatic text retrieval. *Science*, 253:974–979, 1991.

[185] T. Sato, T. Kanade, E. Hughes, and M. Smith. Video OCR for digital news archive. In *Workshop on Content-Based Access of Image and Video Databases*, pages 52–60, Los Alamitos, CA, January 1998.

[186] R. Schapire. Strength of weak learnability. *Journal of Machine Learning*, 5:197–227, 1990.

[187] B. Schölkopf and A. Smola. *Learning with Kernels Support Vector Machines, Regularization, Optimization and Beyond*. MIT Press, Cambridge, MA, 2002.

[188] C. Shannon. Prediction and entropy of printed English. *Bell Sys. Tech. Journal*, 30:50–64, 1951.

[189] V. Sindhwani, P. Niyogi, and M. Belkin. Beyond the point cloud: from transductive to semi-supervised learning. In *Proc. ICML*, 2005.

[190] R. Singh, M. L. Seltzer, B. Raj, and R. M. Stern. Speech in noisy environments: Robust automatic segmentation, feature extraction, and hypothesis combination. In *IEEE Conference on Acoustics, Speech, and Signal Processing*, Salt Lake City, UT, May 2001.

[191] J. Sivic, B. Russell, A. Efros, A. Zisserman, and W. Freeman. Discovering object categories in image collections. In *Proc. ICCV*, 2005.

[192] A. W. M. Smeulders, M. Worring, S. Santini, A. Gupta, and R. Jain. Content-based image retrieval at the end of the early years. *IEEE Trans. on Pattern Analysis and Machine Intelligence*, 22:1349–1380, 2000.

[193] J.F. Sowa. *Conceptual Structures: Information Processing in Mind and Machine*. Addison-Wesley, 1984.

[194] J.F. Sowa. *Knowledge Representation — Logical, Philosophical, and Computational Foundations*. Thomson Learning Publishers, 2000.

[195] P. Spirtes, C. Glymour, and R. Scheines. *Causation, Prediction, and Search*. Springer Verlag, New York, 1993.

[196] R.K. Srihari and Z. Zhang. Show&tell: A multimedia system for semi-automated image annotation. *IEEE Multimedia*, 7(3):61–71, 2000.

[197] R. Steinmetz and K. Nahrstedt. *Multimedia Fundamentals — Media Coding and Content Processing*. Prentice-Hall PTR, 2002.

[198] V.S. Subrahmanian. *Principles of Multimedia Database Systems*. Morgan Kaufmann, 1998.

[199] S.L. Tanimoto. *Elements of Artificial Intelligence Using Common LISP*. Computer Science Press, 1990.

[200] B. Taskar, V. Chatalbashev, D. Koller, and C. Guestrin. Learning structured prediction models: A large margin approach. In *Proc. ICML*, Bonn, Germany, August 2005.

[201] B. Taskar, C. Guestrin, and D. Koller. Max-margin Markov networks. In *Neural Information Processing Systems Conference*, 2003.

[202] G. Taubin and D. B. Cooper. Recognition and positioning of rigid objects using algebraic moment invariants. In *SPIE: Geometric Methods in Computer Vision Proceedings*, volume 1570, pages 175–186, 1991.

[203] Y.W. Teh, M.I. Jordan, M.J. Beal, and D.M. Blei. Hierarchical Dirichlet process. *Journal of the American Statistical Association*, 2006.

[204] B.T. Truong, S. Venkatesh, and C. Dorai. Automatic genre identification for content-based video categorization. In *International Conference on Pattern Recognition (ICPR)*, Los Alamitos, CA, 2000.

[205] E.P.K. Tsang. *Foundations of Constraint Satisfaction*. Academic Press, 1993.

[206] I. Tsochantaridis, T. Hofmann, T. Joachims, and Y. Altun. Support vector machine learning for interdependent and structured output spaces. In *Proc. ICML*, Banff, Canada, 2004.

[207] V. Vapnik. *The Nature of Statistical Learning Theory*. Springer, New York, 1995.

[208] V. Vapnik and A. Lerner. Pattern recognition using generalized portrait method. *Automation and Remote Control*, 24, 1963.

[209] V.N. Vapnik. *Statistical Learning Theory*. John Wiley & Sons, Inc, 1998.

[210] N. Vasconcelos and A. Lippman. Bayesian relevance feedback for content-based image retrieval. In *IEEE Workshop on Content-based Access of Image and Video Libraries (CBAIVL'00)*, Hilton Head, South Carolina, June 2000.

[211] C. Vertan and N. Boujemaa. Embedding fuzzy logic in content based image retrieval. In *The 19th Int'l Meeting of the North America Fuzzy Information Processing Society Proceedings*, Atlanta, July 2000.

[212] J.Z. Wang, J. Li, and G. Wiederhold. SIMPLIcity: Semantics-sensitive integrated matching for picture libraries. *IEEE Trans. on PAMI*, 23(9), September 2001.

[213] X. Wang and E. Grimson. Spatial latent Dirichlet allocation. In *Proc. NIPS*, 2007.

[214] X. Wang, X. Ma, and E. Grimson. Unsupervised activity perception by hierarchical Bayesian models. In *Proc. CVPR*, 2007.

[215] M.K. Warmuth, J. Liao, and G. Ratsch. Totally corrective boosting algorithms that maximize the margin. In *Proceedings of the International Conference on Machine Learning (ICML)*, 2006.

[216] P.D. Wasserman. *Neural Computing: Theory and Practice*. Coriolis Group, New York, 1989.

[217] T. Westerveld and A. P. de Vries. Experimental evaluation of a generative probabilistic image retrieval model on "easy" data. In *The SIGIR Multimedia Information Retrieval Workshop 2003*, August 2003.

[218] D. H. Widyantoro, T. R. Ioerger, and J. Yen. An adaptive algorithm for learning changes in user interests. In *Proc. CIKM*, 1999.

[219] I.H. Witten, L.C. Manzara, and D. Conklin. Comparing human and computational models of music prediction. *Computer Music Journal*, 18(1):70–80, 1994.

[220] E. Wold, T. Blum, D. Keislar, and J. Wheaton. Content-based classification, search and retrieval of audio. *IEEE Multimedia*, 3(3):27–36, 1996.

[221] M. E. J. Wood, N. W. Campbell, and B. T. Thomas. Iterative refinement by relevance feedback in content-based digital image retrieval. In *ACM Multimedia 98 Proceedings*, Bristol, UK, September 1998.

[222] Y. Wu, E.Y. Chang, K.C.-C. Chang, and J.R. Smith. Optimal multimodal fusion for multimedia data analysis. In *The ACM MM'04*, New York, New York, October 2004.

[223] Y. Wu, B.L. Tseng, and J.R. Smith. Ontology-based multi-classification learning for video concept detection. In *IEEE International Conference on Multimedia and Expo (ICME)*, June 2004.

[224] R. Yan, J. Yang, and A.G. Hauptmann. Learning query-class dependent weights in automatic video retrieval. In *ACM Multimedia*, New York, NY, 2004.

[225] C. Yang and T. Lozano-Perez. Image database retrieval with multiple-instance learning techniques. In *Proc. ICDE*, 2000.

[226] Y. Yang and C.G. Chute. An example-based mapping method for text categorization and retrieval. *ACM Transactions on Information Systems*, 12(3):252–277, 1994.

[227] J. Yao and Z. Zhang. Object detection in aerial imagery based on enhanced semi-supervised learning. In *Proc. ICCV*, 2005.

[228] J. Yao and Z. Zhang. Semi-supervised learning based object detection in aerial imagery. In *Proc. CVPR*, 2005.

[229] H. Yu and W. Wolf. Scenic classification methods for image and video databases. In *SPIE International Conference on Digital Image Storage and Archiving Systems*, volume 2606, pages 363–371, 1995.

[230] K. Yu, W.-Y. Ma, V. Tresp, Z. Xu, X. He, H.-J. Zhang, and H.-P. Kriegel. Knowing a tree from the forest: Art image retrieval using a society of profiles. In *ACM MM Multimedia 2003 Proceedings*, Berkeley, CA, November 2003.

[231] L. A. Zadeh. Fuzzy sets. *Information and Control*, 8(3):338–353, 1965.

[232] L. A. Zadeh. Fuzzy orderings. *Information Scineces*, (3):117–200, 1971.

[233] O. Zaiane, S. Smirof, and C. Djeraba, editors. *Knowledge Discovery from Multimedia and Complex Data.* Springer, 2003.

[234] M. Zeidenberg. *Neural Network in Artificial Intelligence.* Ellis Horwood Limited, England, 1990.

[235] H. Zhang, R. Rahmani, S.R. Cholleti, and S.A. Goldman. Local image representations using pruned salient points with applications to CBIR. In *Proc. ACM Multimedia,* 2006.

[236] Q. Zhang and S.A. Goldman. EM-DD: An improved multiple-instance learning technique. In *Proc. NIPS,* 2002.

[237] Q. Zhang, S.A. Goldman, W. Yu, and J.E. Fritts. Content-based image retrieval using multiple instance learning. In *Proc. ICML,* 2002.

[238] R. Zhang, S. Khanzode, and Z. Zhang. Region based alpha-semantics graph driven image retrieval. *Proc. International Conference on Pattern Recognition,* Cambridge, UK, August 2004.

[239] R. Zhang, R. Sarukkai, J.-H. Chow, W. Dai, and Z. Zhang. Joint categorization of queries and clips for Web-based video search. *Proc. International Workshop on Multimedia Information Retrieval,* Santa Barbara, CA, November 2006.

[240] R. Zhang and Z. Zhang. Hidden semantic concept discovery in region based image retrieval. In *IEEE International Conference on Computer Vision and Pattern Recogntion (CVPR) 2004,* Washington, DC, June 2004.

[241] R. Zhang and Z. Zhang. A robust color object analysis approach to efficient image retrieval. *EURASIP Journal on Applied Signal Processing,* 2004(6):871–885, 2004.

[242] R. Zhang and Z. Zhang. Fast: Towards more effective and efficient image retrieval. *ACM Multimedia Systems Journal,* 10(6), October 2005.

[243] R. Zhang and Z. Zhang. Effective image retrieval based on hidden concept discovery in image database. *IEEE Transactions on Image Processing,* 16(2):562–572, 2007.

[244] R. Zhang, Z. Zhang, and S. Khanzode. A data mining approach to modeling relationships among categories in image collection. *Proc. ACM International Conference on Knowledge Discovery and Data Mining,* Seattle, WA, August 2004.

[245] R. Zhang, Z. Zhang, M. Li, W.-Y. Ma, and H.-J. Zhang. A probabilistic semantic model for image annotation and multi-modal image retrieval. In *Proc. IEEE International Conference on Computer Vision,* 2005.

[246] R. Zhang, Z. Zhang, M. Li, W.-Y. Ma, and H.-J. Zhang. A probabilistic semantic model for image annotation and multi-modal image retrieval. *ACM Multimedia Systems Journal*, 12(1):27–33, 2006.

[247] T. Zhang and C.-C. Kuo. Audio content analysis for online audiovisual data segmentation and classification. *IEEE Transactions on Speech and Audio Processing*, 9(4):441–457, 2001.

[248] Z. Zhang, Z. Guo, C. Faloutsos, E.P. Xing, and J.-Y. Pan. On the scalability and adaptability for multimodal retrieval and annotation. In *Proc. International Workshop on Visual and Multimedia Digital Libraries*, Modena, Italy, 2007.

[249] Z. Zhang, R. Jing, and W. Gu. A new Fourier descriptor based on areas (AFD) and its applications in object recognition. In *Proc. of IEEE International Conference on Systems, Man, and Cybernetics*. International Academic Publishers, 1988.

[250] Z. Zhang, F. Masseglia, R. Jain, and A. Del Bimbo. Editorial: Introduction to the special issue on multimedia data mining. *IEEE Transactions on Multimedia*, 10(2):165–166, 2008.

[251] Z. Zhang, R. Zhang, and J. Ohya. Exploiting the cognitive synergy between different media modalities in multimodal information retrieval. In *The IEEE International Conference on Multimedia and Expo (ICME'04)*, Taipei, Taiwan, July 2004.

[252] R. Zhao and W.I. Grosky. Narrowing the semantic gap — improved text-based web document retrieval using visual features. *IEEE Trans. on Multimedia*, 4(2), 2002.

[253] X. S. Zhou, Y. Rui, and T. S. Huang. Water filling: A novel way for image structural feature. In *IEEE Conf. on Image Processing Proceedings*, 1999.

[254] Z.-H. Zhou and J.-M. Xu. On the relation between multi-instance learning and semi-supervised learning. In *Proc. ICML*, 2007.

[255] L. Zhu, A. Rao, and A. Zhang. Theory of keyblock-based image retrieval. *ACM Transaction on Information Systems*, 20(2):224–257, 2002.

[256] Q. Zhu, M.-C. Yeh, and K.-T. Cheng. Multimodal fusion using learned text concepts for image categorization. In *Proc. ACM Multimedia*, 2006.

[257] X. Zhu. Semi-supervised learning literature survey. *Technical Report*, 1530, 2005.

[258] X. Zhu, Z. Ghahramani, and J. Lafferty. Time-sensitive Dirichlet process mixture models. *Technical Report CMU-CALD-05-104*, 2005.

[259] X. Zhu, Z. Ghahramani, and J.D. Lafferty. Semi-supervised learning using Gaussian fields and harmonic functions. In *Proc. ICML*, pages 912–919, 2003.

[260] H. Zimmermann. *Fuzzy Set Theory and Its Applications*. Kluwer Academic Publishers, 2001.

Index